T0146141

Characteristics and Duty Limitations of Service Members Transferring Between the Active and Reserve Components

HEATHER KRULL, CHRISTINA PANIS, KATHERINE ANANIA, PHILIP ARMOUR

Prepared for the Office of the Secretary of Defense
Approved for public release; distribution unlimited

NATIONAL DEFENSE RESEARCH INSTITUTE

For more information on this publication, visit www.rand.org/t/RR3100

Library of Congress Cataloging-in-Publication Data is available for this publication.
ISBN: 978-1-9774-0391-9

Published by the RAND Corporation, Santa Monica, Calif.
© Copyright 2020 RAND Corporation
RAND® is a registered trademark.

Support RAND
Make a tax-deductible charitable contribution at
www.rand.org/giving/contribute

www.rand.org

Preface

In 2014, the Department of Defense (DoD) Inspector General (IG) published a report, *Assessment of DoD-Provided Healthcare for Members of the United States Armed Forces Reserve Components*, that observed that service members were being discharged from the active component (AC) with medical conditions that limited their deployability and then affiliating with the reserve component (RC), thereby decreasing individual medical readiness rates (U.S. Department of Defense Inspector General, 2014). DoD IG recommended that the Under Secretary of Defense for Personnel and Readiness "describe its plan to establish guidance that directs the Services to establish criteria and procedures that will ensure AC service members who transfer into the Selected Reserve meet IMR requirements" (U.S. Department of Defense Inspector General, 2014, p. 21).

This report contains a review of DoD and service policies that define requirements for transfers from the AC to the RC and a description of how those policies are implemented, an analysis of the characteristics of service members who separated from the AC between FY 2010 and FY 2016 and later affiliated with the RC, and an analysis of duty limitations observed among AC to RC transfers. The analysis of duty limitations also includes a retrospective look at what information was available during the service member's time in the AC that was related to the RC medical condition. Finally, the report concludes with a set of recommendations that, if implemented, should reduce the number of service members who transfer from the AC to the RC with medical conditions that limit their deployability.

This research was sponsored by the Office of the Under Secretary of Defense for Personnel and Readiness, Manpower and Reserve Affairs, and conducted within the Forces and Resources Policy Center of the RAND Corporation's National Defense Research Institute, a federally funded research and development center sponsored by the Office of the Secretary of Defense, the Joint Staff, the Unified Combatant Commands, the Navy, the Marine Corps, the defense agencies, and the defense Intelligence Community.

For more information on the RAND Forces and Resources Policy Center, see www.rand.org/nsrd/ndri/centers/frp or contact the director (contact information is provided on the webpage).

Contents

Figures

Tables

Summary

The Department of Defense (DoD) Inspector General (IG) published a report in 2014, *Assessment of DoD-Provided Healthcare for Members of the United States Armed Forces Reserve Components*, with an observation about service members who transfer from the active component (AC) to the reserve component (RC) (Department of Defense Office of the Inspector General, 2014). Specifically, DoD IG observed that medically limited or nondeployable service members were affiliating with the RC (Selected Reserve [SELRES]) after being discharged from the AC, despite policies in place establishing the criteria for transferring, and therefore individual medical readiness (IMR) rates were reduced. DoD IG therefore recommended that the Under Secretary of Defense for Personnel and Readiness (USD[P&R]) develop a plan to establish guidance that charges the services with establishing procedures and criteria that will ensure AC to RC transfers meet IMR requirements.

In 2016, the Office of the Under Secretary of Defense for Personnel and Readiness, Manpower and Reserve Affairs, asked RAND National Defense Research Institute (NDRI) to conduct an analysis to help them identify policy improvements to address DoD IG's concern. RAND NDRI reviewed existing DoD and service policies regarding how service members transfer into the RC, as well as policies about deployability. Then, using individual-level administrative data, RAND NDRI conducted an analysis of the number and characteristics of service members transferring from the AC to the RC, their duty limitations after affiliating with the RC, and how well information from the service member's time in the AC relates to future readiness or deployability issues once the member affiliates with the RC. Based on the results of this analysis, RAND NDRI offers recommendations for how to improve screening with the goal of reducing the number of service members who affiliate with the RC with medical conditions that limit their deployability.

Department of Defense and Service Policies on Active Component to Reserve Component Transfers and Deployability

DoD policy that establishes how service members may transfer into the ready reserve, the category of reservist to which selected reservists belong, cites two sections of Title 10 of the U.S. Code (10 U.S.C.) that deal with the total service required of members in the armed forces (which may be completed in the ready reserve) and how service members are placed in the ready reserve. Neither section specifies screening to help identify medical issues that may result in nondeployability. The DoD instructions that establish medical requirements for enlistment, appointment, or induction into the military exempt service members who affiliate with the reserves soon after being discharged from the AC, as transfers do. Service policies vary in their requirements for AC to RC transfers but generally require service members who affiliate with the RC within six months of separating from the AC to meet medical retention standards or to undergo a review of their medical history, with additional screening done for new or changed conditions since discharge from the AC. The services use duty-limitation information, Department of Veterans Affairs (VA) disability ratings, health records, and/or results of the Separation History and Physical Examination (SHPE) in their assessment of the service member's qualification for transfer.

DoD's IMR policy, referenced by DoD IG, defines six elements of IMR: periodic health assessment, deployment-limiting medical and dental conditions, dental assessment, immunization status, medical readiness and laboratory studies, and individual medical equipment. Our analysis focuses on the second element, duty limitations, which serve as one way to measure deployability. DoD Instruction (DoDI) 1332.45, *Retention Determinations for Non-deployable Service Members* (Department of Defense Instruction 1332.45, 2018b), specifies that services members with medical conditions that require light duty for more than 30 days are considered temporarily nondeployable.

Characteristics and Duty Limitations of Active Component to Reserve Component Transfers

Using individual-level administrative data from a variety of sources, we created an analytic file of service members who were discharged from the AC between FY 2010 and FY 2016 and affiliated with the RC within 24 months of discharge. We found that 20,000 to 25,000 service members transferred from the AC to the RC each year (a total of just more than 130,000 over 7 years), over 80 percent of whom transferred within 6 months of separation. Because it is these transfers who are subject to reduced or no medical screening in order to affiliate with the RC, they are the focus of the remainder of our analysis.

Among transfers, 80 percent join the RC within 1 to 2 months of leaving the AC. Most affiliate into the same RC service as the AC service they left, with slightly more transfers joining the Army and Air Force RCs than the number who served in these services while in the AC. More than 85 percent of transfers were enlisted, and over 40 percent had 5 or fewer years of service at the time of transfer. This is not surprising given that all service members have an obligation to serve for 6 to 8 years, and those who choose to not serve all of that time in the AC must complete their service in an RC (10 U.S.C. 651, 2017). Officers transferred later in their careers than enlisted personnel. Approximately 60 percent of transfers were observed for at least 24 months in the RC. Among those who left in the first 24 months, between 2 and 5 percent returned to the AC. A much larger percentage, 44 percent overall, transferred to the individual ready reserve (IRR).

We received AC and RC duty-limitation (profile) data for the Army and Air Force (but not for the Navy or Marine Corps). Approximately 30 percent of transfers in both services were observed with an RC profile, more than 75 percent of which were 30 or more days in duration. Per DoDI 1332.45 (Department of Defense Instruction 1332.45, 2018b), described above, that means one of out of every four to five Army and Air Force transfers were temporarily nondeployable after joining the RC. We compared the start date of the profile to the date of RC affiliation and found that one-third of transfers in the Army were on an RC profile within the first six months after affiliation. The most common conditions on RC profiles for soldiers were musculoskeletal (nontraumatic joint disorders, such as those affecting the knee, hip, shoulder, or ankle, and back problems), anxiety disorders (usually post-traumatic stress disorder [PTSD]), and hearing loss. Among airmen, pregnancy was listed on nearly one-third of all RC profiles. Like soldiers, nontraumatic joint disorders and back problems were also common, as were disorders of the teeth and jaw.

We used AC profile conditions and conditions identified during the service member's SHPE to assess the extent to which RC profile conditions appeared earlier, during the individual's AC service. Approximately one-third of soldiers with RC profiles for nontraumatic joint disorders and back problems had a similar condition listed on their SHPE. One-quarter of soldiers with RC profiles for back problems and hearing disorders also had AC profiles for the same reason. Airmen who were temporarily nondeployable after joining the RC for nontraumatic joint disorders, back problems, other connective tissue disease, and sprains and strains had similar problems in the AC, with AC-RC profile match rates of 40–50 percent for each condition.

Recommendations

The findings of the analyses above yielded four recommendations for improving the screening for AC to RC transfers with the goal of retaining service members free of medical conditions that limit their deployability.

First, DoD should require service members to meet retention standards to be able to affiliate with SELRES. Current DoD policies pertaining to medical requirements for appointment, enlistment, or induction exempt service members who affiliate with the RC within six months of separating from the AC, and the policy that defines procedures for assigning service members to the RC does not address medical requirements. Therefore, currently, DoD does not have a standard for medical requirements for transfers, so we recommend that DoD impose the standard that transfers must meet the receiving service's retention standards.[1] DoD is developing medical retention standards, and they should become the new minimum requirement once they are implemented.[2]

Second, because of the timing of the SHPE and the match rate between conditions identified during that exam and conditions that later appear on RC profiles, we recommend that medical retention standards be applied at the time of the SHPE. Implementing this recommendation may require additional resources or procedural changes. VA conducts some separation exams and may not currently be equipped to evaluate whether a service member meets medical retention standards. Furthermore, since the current purpose of the SHPE is to aid in the transfer of care from the DoD to the VA and to facilitate disability benefits determinations, using the SHPE to evaluate whether the service member meets medical retention standards would require a policy change and adoption of an expansion of the intended scope of the exam, and with it, the need for additional resources.

Third, we recommend that, to the extent possible, the requirements for transferring from the AC to the RC and the information used to determine whether the service member meets these requirements should be standardized across services and components. Current

---

[1] Since this research was completed and the report written, DoDI 1200.15 has been revised to say that service members re-accessing into the ready reserve within 12 months of separation must be fully medically ready according to individual medical readiness standards, and that service members who do and who are not deployable must be approved for retention by the gaining service. See Department of Defense Instruction 1200.15, rev. *Assignment to and Transfer Between Reserve Categories and Discharge from Reserve Status*, Washington, D.C.: U.S. Department of Defense, 2019.

[2] Since this report was written, DoD has published medical retention standards. See Department of Defense Instruction 6130.03 V2, *Medical Standards for Military Service: Retention*, Washington, D.C.: U.S. Department of Defense, 2020.

policies differ between DoD and the services and across services. If DoD policy requires that transfers meet medical retention standards (specifically, those under development by the Retention Medical Standards Working Group) at the time of AC separation, that will set a minimum standard for the services. A common set of standards will improve the transfer process for those leaving one service and joining another and will enable a standardized process for evaluating a service member's candidacy for transfer and what information is used in the assessment.

Finally, we recommend that the AC to RC decision authority have access to and make full use of information available at the time of AC separation, including duty-limiting conditions, VA disability ratings, results of the SHPE, and medical records. Our analysis of the match rate between RC profile conditions and conditions found during SHPE exams and on AC profiles showed a high correspondence for several types of medical issues, including musculoskeletal (nontraumatic joint disorders, back problems, and sprains and strains), some mental health conditions, and hearing impairment. Although the presence of medical problems in the AC may be indicative of future RC duty limitations, screening out too many transfer candidates because of the medical issues during active service would make recruiting to the RC more difficult, especially when not all potential transfers experience duty limitations after affiliating with the RC. It is for this reason that we recommend all available information be used, not just the presence of an AC profile or a medical issue identified during the SHPE. Our analysis found that the vast majority of service members who had an AC profile or a condition identified during the SHPE did *not* have an RC profile for the same condition.

Our analysis showed that one out of every four or five Army and Air Force transfers were temporarily nondeployable within 24 months of affiliating with the RC, the majority in the first several months of RC service. Defining the medical standards that must be met and utilizing information already available are the first steps to reducing the number of service members who transfer to the RC with conditions that limit their deployability.

Acknowledgments

This study was originally sponsored by Stephanie Barna, then Principal Deputy Assistant Secretary of Defense for Manpower and Reserve Affairs. Later, Lernes Hebert served as study sponsor in his role as Principal Director, Military Personnel Policy. Our action officers throughout the course of the study provided valuable support and guidance: Col Janet Pouncy, Office of the Secretary of Defense for Reserve Affairs, Manpower and Personnel; COL George Barido, Office of the Assistant Secretary of Defense for Health Affairs, Reserve Component Medical Programs and Policy; COL Dennis Ratliff, Office of the Assistant Secretary of Defense for Health Affairs, Reserve Component Medical Programs and Policy; and CAPT Sean McDonald, Office of the Assistant Secretary of Defense for Manpower and Reserve Affairs, Military Personnel Policy/Officer and Enlisted Personnel Management.

At the beginning of the study, we met with representatives from the RC to hear their perspectives on DoD IG's observations about medically limited or nondeployable service members transferring to the RC, about service policies that guide the AC to RC transfer process, and about ways those policies and procedures could be improved. We learned a great deal about the process from these conversations, and their insights shaped our approach to the research questions. The analyses in this study required data from a variety of sources: and we appreciate the assistance from individuals at the Defense Manpower Data Center, Defense Health Agency, and Health Policy Services and Oversight who worked to ensure we received the data we needed.

Gail Fisher, then at RAND, participated in the early stages of this study and was instrumental in identifying relevant policies and taking part in discussions with representatives from the RCs.

Finally, we appreciate feedback from Ellen Pint at RAND and James Bishop at the Institute for Defense Analyses on an earlier draft of this report. Their comments improved it substantially.

Abbreviations

AC	active component
ADMF	Active Duty Master File
AFI	Air Force Instruction
AFSC	Air Force Specialty Code
AOC	area of concentration
AR	Army Regulation
ASIMS	Aeromedical Services Information Management System
CCS	Clinical Classification Software
COMNAVCRUITCOMINST	Commander, Navy Recruiting Command Instruction
DES	Disability Evaluation System
DHA	Defense Health Agency
DMDC	Defense Manpower Data Center
DoD	Department of Defense
DoDI	Department of Defense Instruction
ICD	International Classification of Diseases
IDES	Integrated Disability Evaluation System
IG	Inspector General
IMR	individual medical readiness
IRR	individual ready reserve
LIMDU	limited-duty (adj.)
MOS	military occupational specialty

MSO	military service obligation
NAVET	naval veteran
NAVMED	Navy Medicine
PHA	periodic health assessment
PTSD	post-traumatic stress disorder
RC	reserve component
RCCPDS	Reserve Component Personnel Data System
SELRES	Selected Reserve
SHA	Separation Health Assessment
SHPE	Separation History and Physical Examination
U.S.C.	U.S. Code
USD(P&R)	Under Secretary of Defense for Personnel and Readiness
VA	Department of Veterans Affairs
VTA	Veterans Tracking Application
WEX	Work Experience

CHAPTER ONE

Introduction

Background

An October 2014 U.S. Department of Defense (DoD) Inspector General (IG) report, *Assessment of DoD-Provided Healthcare for Members of the United States Armed Forces Reserve Components*, stated,

> Active Component (AC) service members transferred into the Selected Reserve with medical conditions which limited their deployability or for which they were subsequently found non-deployable.
>
> This occurred because established medical entrance criteria used by the Services did not prevent the transferring of medically-limited or non-deployable AC service members to the Selected Reserve.
>
> As a result, medically limited or non-deployable AC service members transferred to the Selected Reserve, which decreased Individual Medical Readiness (IMR) rates. (U.S. Department of Defense Inspector General, 2014, p. 17)

The report went on to say that the services had established medical criteria for service members transferring between the active component and the Selected Reserve (SELRES), but that despite those policies, "medically-limited or non-deployable AC service members have been allowed to transfer into Selected Reserve units" (U.S. Department of Defense Inspector General, 2014, p. 19). While the report did not include individual medical readiness (IMR) rates, it did include observations made by reserve component (RC) unit commanders who expressed concerns about active component (AC) service members with deployment-limiting conditions being transferred to the RC and being subsequently medically not ready.

Based on these findings, DoD IG recommended that the Under Secretary of Defense for Personnel and Readiness (USD[P&R]) "describe its plan to establish guidance that directs the Services to establish criteria and procedures that will ensure AC service members who transfer into the Selected Reserve meet IMR requirements" (U.S. Department of Defense Inspector General, 2014, p. 21).

To address DoD IG's recommendation, the Office of the Under Secretary for Personnel and Readiness, Manpower and Reserve Affairs, asked the RAND Corporation to conduct an analysis to determine the scope of the problem and to identify policy solutions to reduce the number of service members with deployment-limiting medical conditions joining the RC.

Study Objectives

To help USD(P&R) identify policy improvements to address DoD IG's concern, this research study had two aims: to document how DoD and service policies are implemented when service members separate from the AC and plan to join the RC, and to assess and quantify DoD IG's observation that medically limited or nondeployable AC service members were transferring into SELRES. DoD IG referenced but did not report IMR rates, and we did not have access to those data for this study. However, we obtained other data related to duty limitations and medical nonreadiness from the Army and Air Force and report those findings in Chapter Four.[1] We also examined other, more extreme health-related outcomes, including receipt of incapacitation pay and placement on medical orders, as well as medical discharges through

---

[1] We requested but were unable to receive similar data for Navy and Marine Corps personnel.

the Disability Evaluation System (DES). To that end, we developed an analytic file to answer the following questions:

- How many service members transfer from the AC to the RC each year, and what are their characteristics?
- What medical conditions are present prior to AC separation that might reappear after RC affiliation if the service member is permitted to transfer from the AC to the RC?
- How many transfers have a deployment-limiting medical condition after joining the RC, and what are the most common conditions that present?
- What percentage of service members with a deployment-limiting medical condition after RC affiliation had a related condition prior to AC separation?

Approach

To achieve these objectives, we began by reviewing DoD and service policies that define who is eligible to affiliate with the RC following discharge from the AC, the requirements that service members must fulfill in order to be able to transfer, and policies related to deployment qualifications. We met with DoD and service representatives to learn how these policies are implemented and what information is available and used to screen for RC eligibility.

We then built an individual-level analytic file comprised of personnel and pay files, disability records, medical encounters, and duty limitations. We conducted a descriptive analysis of the transfer population (SELRES only), guided by the research questions listed above.

Organization of This Report

Chapter Two describes the DoD and service policies that guide AC to RC transfers and those that determine deployability requirements, as well as the service implementation of these policies. Chapter Three describes the characteristics of the AC to RC transfer population. Chapter Four explores duty limitations and other indicators of medical nonreadiness among the AC to RC transfer population. Finally, Chapter Five summarizes the results of our analysis, acknowledges limitations, and lays out a set of recommendations to reduce the number of service members transferring into the RC with duty limitations that make them nondeployable.

CHAPTER TWO

Policy Review

To give context to what motivated our approach to the analysis in this study, we describe in this chapter DoD-wide and service-specific policies that define how service members may transfer from the AC to the RC. We will also summarize insights from conversations we had with stakeholders about the process of transferring a service member from the AC to the RC and the extent to which they shared DoD IG's concern about medical issues present in AC to RC transfers. We begin by describing why a service member might choose to transfer to the RC after serving in the AC and then describe DoD and service policies that govern how and which service members may transfer from the AC to the RC.

Transferring from the Active Component to Reserve Component

According to U.S. Code, Title 10, Section 651 (10 U.S.C. 651, 2017), individuals who join the military are required to serve for a period of between six and eight years. If the member does not spend all of that time on active duty, the requirement may be fulfilled by serving in an RC. DoD Instruction (DoDI) 1304.25, *Fulfilling the Military Service Obligation (MSO)*, further specifies that "every person who enters military service by enlistment or appointment incurs an MSO of 8 years from that entry date," with three exceptions: (1) enlistments in the National Guard and Reserve are continued until six months after a war or emergency; (2) if the service member is appointed as a commissioned officer in a critically short health professional specialty, the minimum period of obligated service is either two years or the agreed-upon terms for receipt of an accession bonus or special pay; and (3) if the service member leaves the AC and directly affiliates with SELRES, the service member receives a waiver to reduce the military service obligation (MSO) from eight years to six years (Department of Defense Instruction 1304.25, 2013b).

When an individual joins the AC, they agree to a term of service that is usually less than the MSO. The length of the term depends on service, occupation, the recruiting environment, and potentially other factors.[1] Those eligible for reenlistment may complete their MSO by serving additional time in the AC, but those who choose to leave the AC must fulfill their obligation in an RC, typically in the individual ready reserve (IRR) or SELRES. Service members in SELRES more actively participate in military training activities and are typically the first to be activated (Department of Defense Instruction 1235.13, 2013a; Department of Defense Instruction 1215.13, 2015b; *Military Times*, undated). When a service member chooses to leave the AC, they participate in counseling, where they receive "an explanation of the procedures for and advantages of affiliating with the Selected Reserve" (10 U.S. Code 1142, 2017). Joining SELRES may come with incentives, such as bonuses (see, e.g., Sanborn, 2014). Service members who choose to transfer to SELRES must either match a billet in the local unit, wait until a billet becomes available, join a different unit with an open billet, or retrain to a new occupation that has an open billet (Department of Defense Instruction 1215.13, 2015b).

We now turn to DoD and service policies that govern how and which service members may transfer from the AC to the RC. Table 2.1 summarizes the highlights of these policies as they relate to this study's research questions.

---

[1]   When recruiting is challenging or the recruiting mission changes, the services may offer short contracts to fill the mission. See, for example, Myers, 2017.

Table 2.1
Summary of Key Department of Defense and Service Policy Documents

Policy	Title	Relevance to AC to RC Transfers
DoDI 1200.15	*Assignment to and Transfer Between Reserve Categories, Discharge from Reserve Status, Transfer to the Retired Reserve, and Notification of Eligibility for Retired Pay*	Defines procedures for assignment to reserve categories but does not define medical standards
DoDI 6130.03	*Medical Standards for Appointment, Enlistment, or Induction in the Military Services*	Defines accession medical standards, which are not applicable to service members whose separation physical occurred within the last 12 months
DoDI 1304.26	*Qualification Standards for Enlistment, Appointment, and Induction*	Sets policies and responsibilities for entrance qualification standards, including standards for medical and physical fitness, but exempts applicants who apply for reenlistment into the RC following a gap of fewer than six months since discharge from the AC
Army Regulation 601-280	*Army Retention Program*	2006 version stated, "Soldiers fully eligible to reenlist in the Regular Army based on their last physical examination are qualified to join an ARNGUS or USAR unit with regard to the date of the last physical examination"; 2016 version requires a physical examination within 12 months of separation from active duty, validated within 30 days of separation, and soldiers not qualified to reenlist are not eligible to join the RC
Air Force Instruction 48-123	*Medical Examinations and Standards*	Specifies that service members who have separated or retired from active duty and wish to reenlist in the RC must meet standards for continued military service (retention) if no more than six months have elapsed between separation and reenlistment; defines other requirements for reenlistment, including recency of periodic health assessment (PHA) and documentation of retention qualifications

Table 2.1—Continued

Policy	Title	Relevance to AC to RC Transfers
Commander, Navy Recruiting Command Instruction 1130.8J	*Navy Recruiting Manual—Enlisted*	States that Navy veterans who are separated fewer than six months and were physically qualified for separation, are not changing rate or designator (occupation), have no factors that limit worldwide assignment or deployability, do not have a pending VA disability compensation, and had one of four separation codes do not require additional medical review
Navy Recruiting Command Instruction 1131.2F	*Navy Recruiting Manual—Officer*	Confirms enlisted requirements in Commander, Navy Recruiting Command Instruction, for enlisted personnel who separated within six months of RC affiliation
DoDI 6025.19	*Individual Medical Readiness*	Defines elements of individual medical readiness to be PHA, deployment-limiting medical and dental conditions, dental assessment, immunization status, medical readiness and laboratory studies, and individual medical equipment; defines "a deployment-limiting medical condition includes any physical or psychological condition that may interfere with the service member's ability to perform duties while deployed" (Department of Defense Instruction 6025.19, 2014b, p. 9)
DoD 1332.45	*Retention Determinations for Non-deployable Service Members*	Establishes definitions for temporary nondeployability and permanent nondeployability

Department of Defense–wide Policies Governing Active Component to Reserve Component Transfers

DoDI 1200.15, *Assignment to and Transfer Between Reserve Categories, Discharge from Reserve Status, Transfer to the Retired Reserve, and Notification of Eligibility for Retired Pay*, provides procedures for assigning individuals to and transferring between reserve categories, including the ready reserve, to which selected reservists belong (Department of Defense Instruction 1200.15, 2014a). The section on accession into the ready reserves includes a category for service members transferring from active duty and cites two sections, 651 and 10145, of 10 U.S.C. Section 651 describes service obligations of members of the armed forces and how that commitment may be fulfilled, and Section 10145 describes how members may be placed in the ready reserve. This instruction does not define accession or retention medical standards or screening that should be done to ensure AC service members will be healthy enough to deploy on assignment to or transfer into the RC.

DoDI 6130.03, *Medical Standards for Appointment, Enlistment, or Induction in the Military Services*, sets the accession medical standards for military service (Department of Defense Instruction 6130.03, 2018a). Organized by areas of the body or function (e.g., head, ears, hearing), the instruction lists disqualifying conditions, which serve, in part, to ensure that service members are "free of medical conditions or physical defects that may reasonably be expected to require excessive time lost from duty for necessary treatment or hospitalization, or may result in separation from the Military Service for medical unfitness" (Department of Defense Instruction 6130.03, 2018a, p. 4). If a service member cannot meet the standards defined in this DoDI, he or she faces dismissal or rejection from the service. This instruction applies to "applicants for re-accession in Regular and Reserve Components and in federally recognized units or organizations of the National Guard after a period of more than 12 months have elapsed since the separation physical" (Department of Defense Instruction 6130.03, 2018a, p. 9).

DoDI 1304.26, *Qualification Standards for Enlistment, Appointment, and Induction*, delineates policies and responsibilities for entrance qualification standards, including standards for medical and physical fitness (Department of Defense Instruction 1304.26, 2015a). The instruction applies to, among other candidates, "applicants for reenlistment following release from active duty into subsequent Regular or Reserve Components (including the Army National Guard of the United States and the Air National Guard of the United States) after a period of more than 6 months has elapsed since discharge" (Department of Defense Instruction 1304.26, 2015a, p. 2) In other words, the entrance qualification standards that apply to most candidates for military service (e.g., initial enlistment, appointment as commissioned or warrant officers, applicants to the Reserve Officer Training Corps, and the general category "all individuals being inducted into the Military Services" [Department of Defense Instruction 1304.26, 2015a,

p. 2]) do not apply to those who are discharged from one component and reenlist with another within six months. In consultation with the study sponsor, we elected to use a six-month gap between AC separation and RC affiliation as our definition of a transfer. It is these service members, not subject to medical and fitness standards, who are perhaps most at risk of becoming nondeployable in the way DoD IG observed. Furthermore, DoDI 6130.03 uses the separation physical as the anchor for determining who is subject to the standards (Department of Defense Instruction 6130.03, 2018a), whereas DoDI 1304.26 (Department of Defense Instruction 1304.26, 2015a) uses the date of discharge, a date that is more readily available in administrative data used in later chapters in this report.

At the time this research was conducted, DoD did not have retention standards.[2] DoDI 6130.03 (V1 in the 2020 update) and 1304.26 apply to service members accessioning into the military, either for the first time or after being out of the armed services for a period of time. Each individual service has accession and retention standards; everyone interested in joining the military must adhere to DoD accession standards as well as any service standards and policies (which must be at least as strict as DoD's).

In addition to the DoD Instructions described above, each of the services has established policies for the transfer of service members from the AC to the RC.

Service Policies Governing Active Component to Reserve Component Transfers

Army

Army Regulation (AR) 601-280, *Army Retention Program*, has a chapter dedicated to the enlistment and/or transfer processing of soldiers from the regular army to the RC (Chapter Seven) (Army Regulation 601-280, 2016). Section 4 of this chapter defines eligibility for process-

---

[2] Since this report was written, DoD has published medical retention standards. See Department of Defense Instruction 6130.03 V2, *Medical Standards for Military Service: Retention*, Washington, D.C.: U.S. Department of Defense, 2020.

ing into the Army National Guard, U.S. Army Reserve, and the IRR. Among other criteria such as age, Section 7-4 of the 2016 version of AR 601-280 states that "soldiers not qualified to reenlist due to medical and physical fitness criteria contained in paragraph 3-8d are not eligible to join a RC" (Army Regulation 601-280, 2016, p. 32). The regulation further states that "soldiers must have a physical examination completed within 12 months of separation of active duty, and it must be validated within 30 days of transition from active duty" (Army Regulation 601-280, 2016, p. 32). The 2006 version of the AR 601-280 section on medical eligibility for transfer phrased the overall requirement differently: "Soldiers fully eligible to reenlist in the Regular Army based on their last physical examination are qualified to join an ARNGUS or USAR unit with regard to the date of the last physical examination" (Army Regulation 601-280, 2006, p. 38). The analysis in the chapters that follow in this report will be of service members who separated from the AC between FY 2010 and FY 2016. Therefore, the 2006 version of AR 601-280, which did not specify how recently a physical examination need to have occurred, would have guided these transfers.

Air Force

The main document that governs Air Force medical requirements for transfers is Air Force Instruction (AFI) 48-123, *Medical Examinations and Standards*, the most recent version of which was published in August 2013 (Director of Medical Operations and Research, 2013). Throughout, this instruction differentiates between requirements for those who have had a break in service of more than six months from those whose break is shorter. For example, Chapter Five, on continued military service (retention standards), applies to "all individuals who have separated or retired from [active duty] AD with any of the regular Armed Services, but who are reenlisting in the regular Air Force or ARC [Air Reserve component] when no more than 6 months have elapsed between separation and reenlistment. If more than 6 months have elapsed Chapter 4 applies" (Director of Medical Operations and Research, 2013, p. 24), where Chapter Four defines the medical standards for appointment, enlistment, and induction (i.e., accession).

Chapter Seven of this instruction, which specifies when medical examinations are required for separation and retirement, explains that a medical examination is mandatory when the service member has not had a PHA within 1 year. Further, if the service member is transferring to an Air Reserve component, their retention qualifications must be documented, and those who have had a PHA within 12 months prior to the date of separation or retirement will complete a self-assessment within 180 days of transfer, which may result in additional screening, particularly for those with an extensive medical history and/or new signs or symptoms of medical problems. Finally, airmen who are separating early from active duty and being assigned to an Air Reserve component under PALACE CHASE or PALACE FRONT and whose most recent PHA was completed more than 12 months prior must have a medical examination done (Air National Guard, undated).[3]

AFI 48-123 was updated in 2013 from a 2009 version (Air Force Instruction 48-123, 2009); no substantial changes related to medical requirements of the RC were made during this revision. Therefore, the airmen we analyze in the following chapters were all governed by the above policy at the time of transfer.

Navy and Marine Corps

The Navy has a recruiting manual for enlisted personnel, Commander, Navy Recruiting Command Instruction (COMNAVCRUIT-COMINST) 1130.8J, *Navy Recruiting Manual—Enlisted*, that specifies when naval veterans (NAVETs) do and do not require medical review prior to affiliating with the RC (Commander, Navy Recruiting Command, 2011). Those who are separated fewer than six months and were physically qualified for separation, are not changing rate or designator (occupation), have no factors that limit worldwide assignment or deployability, do not have a pending VA disability compensation, and

---

[3] PALACE CHASE is a program that allows airmen who have completed at least half of their original contract (two-thirds for officers) to transition from the Air Force to the Air Guard. PALACE FRONT allows airmen who have completed their active-duty service obligation term to transfer from the Air Force into the Air Guard without a break in service as long as Air Guard eligibility requirements are met.

were separated with one of four codes (RE-R1, RE-1, RE-1E, or RE-6) do not require additional medical review. Pregnant NAVETs with a physical exam completed in the past 24 months may affiliate until the thirty-second week of pregnancy.

NAVETs who separated within 24 months and who have a contract do not need a new physical exam provided their paperwork has been properly filed; if more than 90 days has passed since the most recent exam, the NAVET must have a self-assessment form reviewed by a credentialed military provider, highlighting any changes in health. NAVETs who do not have a contract and who have been separated more than six months require a new physical examination.

A second manual, COMNAVCRUITCOMINST 1131.2F, *Navy Recruiting Manual—Officer*, is undated and confirms the requirements listed above for NAVETs who have separated within six months of affiliation. No other medical requirements are mentioned.

The Navy Medicine (NAVMED) Manual of the Medical Department (NAVMED, 2016), provides instruction for enlisted personnel and commissioned officers wishing to affiliate with the Naval and Marine Corps Selected Reserve. Those who separated from naval active duty service within the previous six months and who did not separate for a medical condition will have an evaluation that consists of

- a report of medical history, reviewed by an examiner and including comments on new or changed medical conditions
- a physical exam and laboratory tests for new or changed conditions
- a review of the service member's certificate of release or discharge from active duty (DD Form 214).

If the applicant has no new or changed conditions since separating from active duty, the applicant is considered physically qualified for affiliation with the Naval or Marine Corps Selected Reserve. If a new condition has arisen or materially changed, the service member must undergo a screening of the medical condition(s) according to the standards for enlistment and commissioning outlined in Chapter Three of NAVMED P-117.

The policies described up to this point dictate the requirements for service members who wish to affiliate with the RC after separating from the AC. They are the subject of DoD IG's observation, but ultimately the concern is their deployability after RC affiliation. In the next section, we describe policies pertaining to deployability.

Policies Governing Deployability

DoD IG's observation about AC to RC transfers who had medical conditions that affected their ability to deploy was ultimately about decreased rates of IMR, a concept defined in DoDI 6025.19, *Individual Medical Readiness* (Department of Defense Instruction 6025.19, 2014b). The instruction states that a service member's medical readiness status is to be considered during every clinical encounter and defines the elements of IMR as the PHA, deployment-limiting medical and dental conditions, dental assessment, immunization status, medical readiness and laboratory studies, and individual medical equipment. As we will describe in Chapter Four, the main source of information about medical nonreadiness in this study was data on duty limitations, which coincides with the second IMR element (deployment-limiting medical [and dental] conditions).

According to IMR policy, "a deployment-limiting medical condition includes any physical or psychological condition that may interfere with the service member's ability to perform duties while deployed" (Department of Defense Instruction 6025.19, 2014b, p. 9), where the conditions are defined in DoDI 6490.07, *Deployment-Limiting Medical Conditions for Service Members and DoD Civilian Employees*, and in military department-specific policies (Department of Defense Instruction 6490.07, 2010). DoDI 6490.07 includes a noncomprehensive list of conditions that will generally preclude service members from participating in a contingency deployment unless a waiver is granted. The medical assessment is done through a predeployment health assessment, medical record review, and a current PHA. This instruction does not specifically pertain to AC to RC transfers.

Although not in place at the time of the DoD IG report, DoDI 1332.45, *Retention Determinations for Non-deployable Service Members*, provides some useful definitions for our assessment of deployability among transfers (Department of Defense Instruction 1332.45, 2018b). The purpose of the policy is to instruct on the handling of service members who are considered nondeployable for more than 12 consecutive months. The temporarily nondeployable medical category includes patients who are hospitalized and expected to return to full duty in fewer than 12 months, service members with medical conditions that limit full duty (i.e., those on light duty for more than 30 days), and service members who are pregnant or in the postpartum phase. The instruction also defines the permanently nondeployable medical category to include service members on permanent limited duty, those currently under disability evaluation, and RC service members who have a permanent duty limitation and are awaiting a determination of whether the impairment was incurred or aggravated in the line of duty. Our analysis in the following chapters relies on duty limitations (including limitations related to pregnancy) and disability evaluations, but we did not have access to hospital records or line-of-duty determinations.

Both the Army and the Navy offer a deferral from involuntary deployments to those members who are transitioning from active duty to an RC; the Air Force does not offer this option. The Army uses the term *stabilization periods* to describe this deferral. The purpose is to allow soldiers to begin a new civilian career or attend school without being interrupted by a deployment. To be guaranteed a stabilization period, the solider must sign up before leaving active duty. All Army Reserves have a two-year stabilization period option, but National Guard participation in the program varies by state (Kotejin, 2008).

The Navy refers to this nondeployable period as a *mobilization deferment,* and if a sailor joins the reserves within 6 months of active-duty service, they receive 2 years of deferment; those joining within 7 to 12 months of leaving active duty receive a 1-year deferment (Owens, 2007).

Insights from the Services About Active Component to Reserve Component Transfers

Early in this research, in late 2016, we met with representatives from each of the RCs to learn their perspectives on DoD IG's observation. In this section, we present some highlights from those discussions, presented generally rather than attempting to attribute issues to one component or another, and because issues present at the time of these discussions may have since evolved.

In general, the services agreed with DoD IG's observation that service members affiliating with the RC after separating from the AC, especially with little or no medical screening occurring prior to joining the RC, often appeared during an initial drill weekend with medical issues. Some described anecdotes of service members bringing a packet of medical information and stating that they were told they were to be medically boarded. We also heard during these discussions that if service members serve for 24 months in the RC, that is viewed as a successful outcome. Already having spent time in the AC, the expectation seemed to be that transfers may not spend much time in the RC, so even 2 years of service is considered valuable. In the empirical analyses that appear in the following chapters, we will examine outcomes separately for transfers who spend at least 24 months in the RC and those who leave in the first 2 years.

The services described varying access to information about transfers. Some or all of the following information was known by the receiving unit: duty-limiting conditions, VA disability ratings, results of the SHPE, and access to medical records. There were concerns about the accuracy of duty-limitation records and about the extent to which some of this information was reviewed, even if available. The presence of a VA rating or pending disability compensation would not preclude a service member from affiliating with SELRES, but it might warrant a more careful review. Some service representatives were also concerned that a stabilization period may induce complacency; not having to deploy for a period of time reduces the incentive to maintain medical readiness.

A recurring theme during these discussions was that a thorough medical examination prior to AC separation would provide an opportunity to evaluate whether the service member meets a set of standards (e.g., accession or retention medical standards).[4] As we will discuss in greater detail in Chapter Four, separating service members who are not filing a disability claim with VA undergo a SHPE. Separating service members are presumed fit for retention unless they have previously been determined unfit and have been serving in a permanent limited-duty status. If the SHPE were instead used to *evaluate* fitness for retention, the exam itself might identify conditions that are likely to reappear once the service member affiliates with the RC.

Service representatives offered other ideas for how to improve the transfer process. One would be to have a medical officer at the local recruiting command do a full review of the service member's health record and determine which medical conditions require a waiver for the service member to be able to affiliate with the RC.[5] At the time of our discussions, there had been some resistance to this option because it would slow down the process of recruiting the service member to the RC; as we'll show in later chapters, the gap between AC separation and RC affiliation is generally very short. There was also discussion about what medical standards would be the right ones to use if the SHPE were used to evaluate a service member's fitness to serve in the RC. Accession medical standards tend to be stricter than retention medical standards; the appropriate degree of strictness might be somewhere in between.

We now turn to our empirical analysis of the characteristics (in Chapter Three) and duty limitations (in Chapter Four) of service members transferring from the AC to the RC.

---

[4]   As mentioned in the previous section, the Army and Air Force apply retention standards to service members affiliating with the RC within six months of separating from AC.

[5]   Note that this is likely not practical for service members who make the decision to transfer to SELRES during separation counseling and therefore do not need to visit a local recruiting station, like someone who has a true break in service would.

CHAPTER THREE

Characteristics of Active Component to Reserve Component Transfers

The first objective of this research was to identify the number and characteristics of service members who, within 6 months of separating from the AC, join the RC. Participation in the RC, for the purposes of this analysis, is restricted to those who join SELRES (either the National Guard or Reserve).[1] Individuals who join the IRR are excluded from analysis. For simplicity, we will refer to affiliation to the RC. To do this, we created an analytic file using personnel data from the Defense Manpower Data Center (DMDC). Specifically, we used DMDC's Work Experience (WEX) file to identify the month and year that service members separated from the AC. We then followed them for 24 months to see if they joined the RC. For those who joined the RC, we continued to follow them to see if any medical issues arose during their first 24 months in the RC, which we describe in the next chapter. This chapter describes the characteristics of the individuals who separated from the AC between FY 2010 and FY 2016 and were later observed in the RC. DMDC's WEX file ends in June 2016, so to create a complete 2016 AC separation cohort, we augmented WEX with the Reserve Component Personnel Data System (RCCPDS) and the Active Duty Master File (ADMF).

---

[1] If a service member transfers from the AC to the National Guard and then to the reserves, we only count the initial transfer out of the AC into the National Guard. We do not count intrareserve transfers as transfers.

Number of Active Component to Reserve Component Transfers

We begin by showing, in Figure 3.1, the number of service members who joined the RC within 24 months of separating from the AC.[2] We divide these service members into three groups: those who join the RC within 6 months of separating from the AC (our main sample), those who join the RC between 7 and 12 months following AC separation, and those who join the RC between 13 and 24 months after separating from the AC.[3]

As the figure shows, there were between 20,000 and 25,000 service members who separated from the AC each year between 2010 and 2016 and who joined the RC within 24 months of separation. The vast majority (83 percent) affiliated with the RC within 6 months of separating from the AC. This same information by service and commissioned officer/enlisted status can be found in Appendix A. The patterns are generally similar, but the percentage of marines affiliating with the RC within the first 6 months is significantly smaller at 54 percent (of all marines who transfer).

Months Between Active Component Separation and Reserve Component Affiliation

For the remainder of our analysis, we restrict the transfer population to those who joined the RC within six months of separating from the AC. We first examine the distribution of months between AC separation and RC affiliation. Figure 3.2 shows that 56 percent of all transfers (all

---

[2]   The number of warrant officer transfers over this period was too small to include in subsequent analyses, so we exclude them from these totals. There were 393 warrant officers across all services who separated from the AC between 2010 and 2016 and joined the RC within 24 months (340 within 6 months, 31 between 7 and 12 months, and another 22 between 13 and 24 months).

[3]   According to policy, a transfer is someone who joins the RC within six months of separating from the AC, and our sample will include only those service members. However, we first performed a sensitivity check to see how many service members join after a six-month gap.

Figure 3.1
Number of Active Component to Reserve Component Transfers,
All Services, 2010–2016

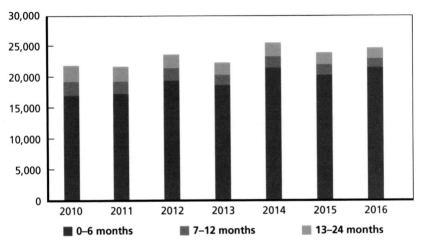

SOURCE: Author calculations using DMDC WEX file.
NOTES: Year is fiscal year of AC separation. Number of months represent months
between AC separation and RC affiliation.

services, enlisted and commissioned officers) have only a one-month
gap between AC separation and RC affiliation.[4] Appendix Figures A.6–
A.9 show some variation by both service and personnel type (enlisted
versus commissioned officers). Across all services, more than half of
commissioned officer transfers affiliate with the RC within one month
of AC separation. However, among enlisted personnel, the distribution
is much smoother among marines and sailors than soldiers and airmen;
60 percent of enlisted soldiers and airmen transfer in the first month,
but only 30–35 percent of marines and sailors do.

---

[4]   The data show that one percent of transfers affiliate with the RC in the same month
they leave the AC. In reality, the distinction between the same month and the next month
is likely an artifact of how WEX is constructed. The dates on records are always the first of
the month, which results in some lost precision. The important takeaway here is that most
transfers affiliate with the RC within two months of separating from AC.

Figure 3.2
Distribution of Gap Between Active Component Separation and Reserve Component Affiliation, All Services, 2010–2016

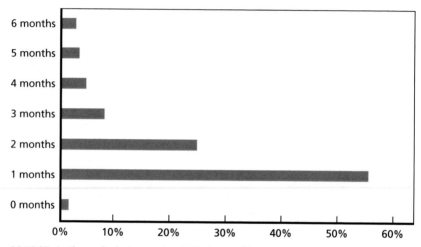

SOURCE: Author calculations using DMDC WEX file.
NOTES: Number of months represent months between AC separation and affiliation with the RC.

Service Affiliation of Active Component to Reserve Component Transfers

Figure 3.3 shows the military service from which the population of transfers separated and which service and component they joined. In general, the number of transfers who left a particular service's AC was similar to the number who joined that service's RC. Slightly more transfers joined the Army RC than the number who left the Army AC, whereas more transfers left the Marine Corps AC than the number who joined the Marine Corps RC.

The split between the reserves and National Guard among Army and Air Force transfers is similar: a larger proportion of transfers who affiliate with the Army and Air Force RC join the reserves than the National Guard (55.5 percent of Army RC transfers and 64.2 percent of Air Force RC transfers join the reserves). To put these numbers into context, Tables B.1–B.3, in Appendix B, show 2013 SELRES

Figure 3.3
Active Component and Reserve Component Service Affiliation Among Transfers, 2010–2016

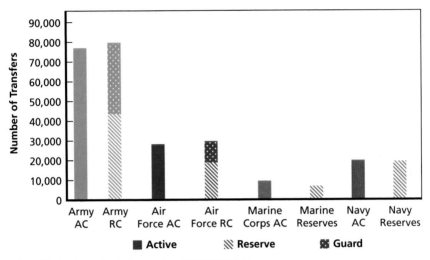

SOURCE: Author calculations using DMDC WEX file.

and active component end strengths by service and 2013 SELRES prior service gains by service.[5] In 2013, the end strength of the Army National Guard was approximately 350,000, and the end strength of the Army Reserve was approximately 200,000. However, the reserve/ National Guard split for all SELRES prior service gains was similar to our results (50.4 percent of Army prior service gains were reserves). The Air National Guard end strength in 2013 was approximately 100,000 compared to 70,000 in the Air Reserve. However, 52.4 percent of Air Force SELRES gains joined the reserves.

Tables B.4–B.5 show interservice transfers. Regardless of whether we examine what services AC separatees transfer to or what service RC

---

[5]   We chose 2013 as a midpoint for our 2010–2016 cohort analysis, even though these numbers may vary over the period studied. Prior service gains is broader than the population we are studying. It can include service members who have more than a six-month gap between the AC and RC, and it can include intrareserve transfers. These tables simply provide context and a general check of our results.

gains transferred from, more than 95 percent of soldiers, airmen, and sailors did not change service. For example, 99.53 percent of Army AC transfers joined the Army National Guard or Army Reserve, and 97.03 percent of transfers into the Army RC served in the AC Army. The exception is the Marine Corps, where only 76.56 percent of Marine Corps AC transfers joined the Marine Reserve. Approximately 18 percent of Marine Corps AC transfers joined the Army RC, 5 percent joined the Air Force RC, and less than 1 percent joined the Naval Reserve. Conversely, nearly 100 percent of transfers into the Marine Corps RC were in the Marine Corps AC.

Pay Grade of Active Component to Reserve Component Transfers

Next, we consider pay grade and years of service of AC to RC transfers, measured as of the date of RC affiliation. As Figure 3.4 shows, 85.7 percent of all transfers are enlisted, with more than half of enlisted transfers E-4s and approximately one-third E-5s. Two-thirds of commissioned officer transfers are O3s. Appendix Figures A.10–A.13 show the same information by service. The Army has the smallest proportion of commissioned officers (8.2 percent of total transfers compared to 19.5 percent in the Air Force, 25.3 percent in the Marine Corps, and 27.1 percent in the Navy). E-4s and E-5s, and O-3s represent the largest proportion of enlisted and commissioned officer transfers, respectively, across services.

Occupation of Active Component to Reserve Component Transfers

In the tables below, we list the ten most frequent occupations among AC to RC transfers, by service and by personnel type (officer and enlisted), at the time of affiliation with the RC. Infantry personnel represented the largest percentage of Army transfers and Marine Corps enlisted transfers. The largest percentage of enlisted and commissioned officer

Figure 3.4
**Pay Grade Distribution of Active Component to Reserve Component
Transfers, All Services, 2010–2016**

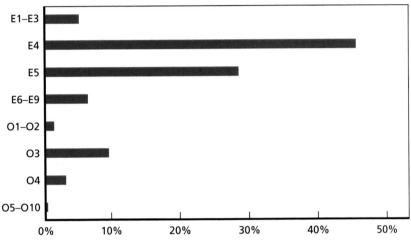

SOURCE: Author calculations using DMDC WEX file.

transfers in the Air Force were aerospace medicine and operations officers, respectively. Pilots/naval flight officers and naval aviators were the
most-represented transfers among Marine Corps and Navy officers,
respectively. Finally, the largest percentage of Navy enlisted transfers
were hospital corpsman, but the distribution was more evenly distrib-
uted across Navy ratings than in the other services. These results generally reflect the overall occupational distribution of the force (Center
for Naval Analyses, 2013).

Years of Service of Active Component to Reserve
Component Transfers

Next, we show years of service at the time of transfer. As Figure 3.5
shows, approximately 40 percent of all transfers have either four or
five years of service when they leave the AC and affiliate with the RC.

Table 3.1
Top-Ten Enlisted and Officer Reserve Component Occupations Among Active Component to Reserve Component Transfers

Enlisted		Commissioned Officer	
Army			
Occupation (Military Occupational Specialty [MOS])	**Percent of Transfers**	**Occupation (Area of Concentration [AOC])**	**Percent of Transfers**
Infantry (11)	16.0	Infantry (11)	8.39
Supply and services (92)	12.6	Military intelligence (35)	7.8
Mechanical maintenance (91)	10.2	Field artillery (13)	6.9
Native language speaker (09)	8.0	Engineer (12)	5.7
Medical (68)	7.3	Nurse corps (66)	5.7
Communications and information systems (25)	5.7	Signal corps (25)	5.1
Transportation (88)	5.5	Armor (19)	4.2
Engineer (12)	5.2	Adjutant general's corps/ ordnance corps (27)	3.8
Field artillery (13)	5.2	Medical service corps (67)	3.5
Armor (19)	4.6	Ordnance corps (91)	3.3
Air Force			
Occupation (Air Force Specialty Code [AFSC])	**Percent of Transfers**	**Occupation (AFSC)**	**Percent of Transfers**
Aerospace maintenance (02A)	20.4	Operations (01)	26.1
Security forces (03P)	12.3	Instructor pilot (K1)	11.3
Cyberspace support (03D)	8.4	Standardization/evaluation airlift pilot (Q1)	10.1
Civil engineering (03E)	7.5	Medical (04)	8.8
Transportation and vehicle management (02T)	5.9	Formal training instructor pilot (T1)	8.2
Munitions and weapons (02W)	5.5	Support (03)	7.0
Intelligence (01N)	4.7	Acquisition (06)	7.0

Table 3.1—Continued

Enlisted		Commissioned Officer	
Command and control systems operations (01C)	3.4	Professional services (05)	5.1
Aerospace Medical Service (04N)	3.1	Logistics (02)	3.8
Materiel Management (02S)	2.9	Weapons and Tactics Instructor Pilot (W1)	2.9
Aircrew Operations (01A)	2.6	Medical Service Specialist (M4)	1.9

Marine Corps

Occupation (MOS)	Percent of Transfers	Occupation (MOS)	Percent of Transfers
Infantry (03)	18.6	Pilots/naval flight officers (75)	22.8
Motor transport (35)	11.0	Infantry (03)	11.9
Communication (06)	10.8	Logistics (04)	10.4
Personnel and administration (01)	9.4	Intelligence (02)	8.6
Supply administration and operations (30)	7.5	Field artillery (08)	7.3
Engineer, construction, facilities, and equipment (13)	4.8	Communication (06)	6.8
Intelligence (02)	3.1	Legal services (44)	5.9
Ground ordnance maintenance (21)	2.9	Supply administration and operations (30)	4.7
Logistics (04)	2.8	Air control/air support/anti-air warfare/air traffic control (72)	4.7
Field artillery (08)	2.7	Miscellaneous (80)	4.0

Navy

Occupation (Rating)	Percent of Transfers	Occupation (Designator)	Percent of Transfers
Hospital corpsman (HM)	9.6	Naval aviator (131)	24.1
Watercraft operator (BM)	6.3	Surface warfare officer (111)	22.0

Table 3.1—Continued

Enlisted		Commissioned Officer	
Master-at-arms/security (MA)	5.7	Submarine warfare officer (112)	9.9
Logistic specialist (LS)	5.6	Naval flight officer (132)	6.9
Yeoman (YN)	4.5	Nurse corps officer (290)	3.8
Information system technician (IT)	3.7	Medical corps officer (210)	3.7
Operation specialist (OS)	3.7	Supply corps officer (310)	3.4
Electronics technician (ET)	3.6	RL officer intelligence officer (183)	2.9
Engineman (EN)	3.2	Judge advocate general's corps officer (250)	2.9
Machinist mate (MM)	3.2	Dental corps officer (220)	1.9

NOTE: We report three-digit occupations, but in the case of 04N, the fourth digit is needed. 04N0 is aerospace medical service, as listed in the table; 04N1 is surgical service. Among the 04Ns reported here, 95 percent are 04N0, so we displayed that label, but it does include some surgical service personnel.

These are likely service members who have completed their active-duty service obligation and fulfill the remainder of their MSO in the RC (10 U.S.C. 651, 2017; Department of Defense Instruction 1304.25, 2013b). Among all transfers, 70 percent have eight or fewer years of service at the time of RC affiliation.

Figures A.14–A.17 repeat the information in Figure 3.5 and show each service's enlisted and officer years-of-service distribution. The Air Force and Navy enlisted years-of-service distribution is similar to the overall distribution among all transfers, with additional upticks in the percentage of transfers at 7 years of service among Air Force enlisted personnel and at 9 years of service among Navy enlisted personnel. The largest percentage of enlisted soldiers transfer in year 4 (28 percent), followed by year 5 (19 percent). The majority of enlisted marines transfer in year 4 (55 percent). Officers tend to transfer later, and the distribution is smoother, especially for transfers from the Marine Corps. Among

Figure 3.5
Years-of-Service Distribution of Active Component to Reserve Component Transfers, All Services, 2010–2016

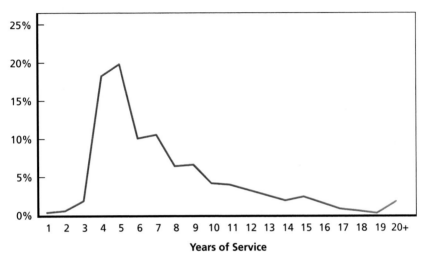

Years of Service

SOURCE: Author calculations using DMDC WEX file.

airmen and sailors, the officer distribution peaks around 11–13 years of service. The largest percentage of Army officers transfer in years 5–8.

Time in the Reserve Component Among Active Component to Reserve Component Transfers

The final characteristic we examine is time spent in the RC after transferring from the AC. As noted in Chapter Two, representatives from the RCs expressed that 24 months of RC service is valuable, so that is how long we follow transfers once they join the RC. Figure 3.6 shows the percentage of transfers in each service who leave the RC in the first 24 months as well as the percentage who stay for at least 24 months. DoD-wide, 60 percent of transfers are observed for at least 24 months in the RC, with marines having the closest split (52.10 percent with 24 or more months, 47.90 leaving in the first 24 months) and airmen having the largest percentage who stay for at least 24 months.

Figure 3.6
Percentage of Transfers Who Serve at Least 24 Months in the Reserve Component and Those Who Serve Fewer Than 24 Months, 2010–2016

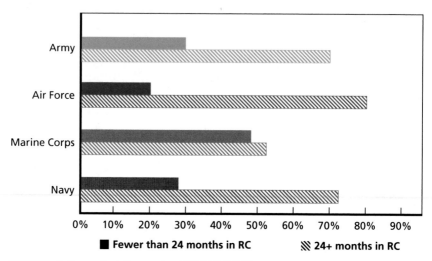

SOURCE: Author calculations using DMDC WEX file.

Summary

Each year, between 20,000 and 25,000 service members affiliate with the RC (SELRES) within 24 months of separating from the active component. The vast majority, 83 percent, transfer in the first 6 months after separating from the AC, and among them, 80 percent transfer within the first 2 months. Enlisted personnel represent 86 percent of service members who transfer to the RC within 6 months of separating from the AC. While there is some movement between services, in general, the number of transfers leaving a service's AC is similar to the number who transfer into that same service's RC. The split between affiliation in the reserves and National Guard in the Army and Air Force is approximately 60 percent Reserve, 40 percent National Guard.

The majority of all transfers are E-4s, followed by E-5s and then O-3s and O-4s, although there is some variation by service, with officers representing a smaller overall share of transfers in the Army than in the other three services. Enlisted marines tend to skew toward fewer years of service at the time of transfer than the other services, but across all services, the largest proportion transfer with 4 or 5 years of service. Officers tend to transfer later in their careers, especially among airmen and sailors, where the largest proportion transfer around year 11 or 12. Once service members transfer, between 50 and 80 percent stay in the RC for at least 2 years, depending on service.

We now turn to duty limitations and other indicators of medical nonreadiness.

CHAPTER FOUR

Duty Limitations and Other Indicators of Medical Nonreadiness

The second and perhaps primary objective of this research was to assess and quantify DoD IG's observation that medically limited or nondeployable AC service members were transferring into SELRES, thereby reducing IMR rates in the RC. DoD IG did not report IMR rates, and we did not have access to those data for this study. However, we obtained other data related to duty limitations and medical nonreadiness and report those findings in this chapter.

As specified in DoDI 6025.19, IMR is comprised of six measurable medical elements: PHA, deployment-limiting medical and dental conditions, dental assessment, immunization status, medical readiness and laboratory studies, and individual medical equipment. We were able to obtain individual-level data on duty-limiting profiles for airmen and soldiers (Department of Defense Instruction 6025.19, 2014b).[1] While not specifically the subject of AC to RC transfers, earlier research examined dental readiness among reservists (Brauner, Jackson, and Gayton, 2012).

In addition to duty-limitation data, we examined other, more extreme health-related outcomes that were present after the service member affiliated with the RC. We obtained data on the receipt of

---

[1] We also requested limited-duty (LIMDU) data from the Department of the Navy, covering marines and sailors transferring between the AC and the RC. At the time of our request, the Navy had recently implemented a new electronic data system for recording information on limited-duty status, and it did not contain enough historical information to enable an analysis of duty limitations among AC to RC transfers.

incapacitation pay and placement on medical orders, as well as medical discharges through DES. While it was beyond the scope of this study to attempt to predict whether a transfer would experience duty limitations or other medical issues after affiliating with the RC, based on information available at the time of separation from the AC, we did obtain some information about medical issues identified during the service member's time in the AC, and we look descriptively at the relationship between this information and RC duty limitations. We begin by looking at duty limitations among soldiers and airmen after joining the RC, the closest measure we have to DoD IG's observation that service members were transferring into the RC with known medical conditions that made them nondeployable.

Reserve Component Duty Limitations

We requested from the Army and Air Force duty-limitation (profile) data on all AC to RC transfers in our sample created using the June 2016 WEX file.[2] We received Air Force profile data covering FY 2007–2018, and the file contained a record for every time an airman was placed on a duty limitation. The information about the profile was restricted to the fiscal year in which it occurred and the duration. The file did not indicate the precise start or end date of the profile, whether it was temporary or permanent, or what limitations were placed on the service member. We received Army profile data covering the period July 2006 through December 2018. Each record indicates the period a service member is on a duty limitation, including the start and end

---

[2]   We also requested LIMDU data from the Navy for airmen and sailors, but we were unsuccessful in obtaining these data primarily because the electronic LIMDU data system had been implemented too recently for there to be sufficient usable data for analysis. The cleanest way for us to construct a sample was using DMDC's June 2016 WEX file. For the analyses in Chapter Three, we added information from RCCPDS and ADMF to create a complete 2016 cohort of transfers, but the files do not merge together seamlessly. Therefore, for the remainder of this report, we rely on only those transfers we could identify through June 2016. The total number of transfers is slightly smaller than reported in Chapter Three (130,852 rather than 136,348).

date of the profile and whether the duty limitation was permanent or temporary. The data also contained information on what body system (e.g., hearing) was impaired and the level of impairment, described in more detail below.

Table 4.1 summarizes the number of transfers for whom we requested profile data and the number of transfers for whom we received profiles. As reported in Chapter Two, our discussions with representatives from the services indicated that 24 months of RC service is one measure of success, so we split Table 4.1 into those transfers who serve for at least 24 months in the RC and those who serve for less time. If medical conditions are associated with less time in service, we might observe higher rates of duty limitations among those who serve for fewer than 24 months. As a benchmark, we also report how many transfers had a profile in the 24 months prior to AC separation and after affiliating with the RC.

Army and Air Force transfers had similar rates of RC duty limitations. Among those who served fewer than 24 months in the RC, approximately 20 percent of transfers had a profile after affiliating with the RC. Among those who served at least 24 months in the RC, approximately one-third were on a duty-limiting profile.[3] These two results suggest that, in the aggregate, it is *not* the case that those who serve fewer than 24 months have a higher rate of duty limitation; therefore, we have no reason to believe that early departure from the RC (fewer than 24 months of service) is the result of medical issues. To the contrary, the higher rate of duty limitations among those who serve at least 24 months is to be expected: the longer someone serves, the more chance they have of experiencing a duty limitation. We explore this issue in more detail below.

The presence of a duty-limiting profile alone is not necessarily an indication of medical nonreadiness. As a point of comparison, we first compare the percentage of transfers who had a profile after affiliating

---

[3] For the remaining 80 percent of service members who stay in the RC for fewer than 24 months and the two-thirds who stay for at least 24 months, we have no profile records for them in our data, implying that in the 2 years we followed them, they were not placed on a duty limitation.

Table 4.1
Number of Transfers with Duty-Limiting Profiles, Army and Air Force

	Army		Air Force	
	Fewer Than 24 Months in the RC	**24+ Months in the RC**	**Fewer Than 24 Months in the RC**	**24+ Months in the RC**
Number of transfers in sample	22,690	53,406	5,750	23,285
Number of transfers with AC profiles in 24 months prior to separation (percent of transfers)	11,279 (49.7%)	28,868 (54.1%)	4,140 (72.0%)	15,261 (65.5%)
Number of transfers with AC profile 30+ days in duration (percent of transfers)	5,288 (23.31%)	13,121 (24.57%)	2,341 (40.71%)	8,395 (36.05%)
Number of transfers with RC profiles (percent of transfers)	5,320 (23.4%)	19,604 (36.7%)	1,086 (18.9%)	7,625 (32.7%)
Number of transfers with an RC P3/P4 profile (percent of transfers)	409 (1.8%)	715 (1.3%)	—	—
Number of transfers with an RC profile 30+ days in duration (percent of transfers)	4,298 (18.9%)	15,342 (28.7%)	827 (14.4%)	5,741 (24.7%)

NOTES: Because we only know fiscal year of profile and duration for airmen, we cannot follow them for exactly 24 months to see if they have a profile. Therefore, following them for 2 years is approximate; for someone who joins the RC in 2012, we look for a profile through 2014. A P3/P4 profile (Army only) indicates a permanent duty limitation of level 3 or 4, where level 3 means the individual has "one or more medical conditions of physical defects that may require significant limitations" and level 4 means the individual has "one or more medical conditions or physical defects of such severity that performance of military duty must be drastically limited" (Army Regulation 40-501, 2017, p. 77).

with the RC to the percentage who had a profile before separating from the AC. Approximately half of all soldiers who transferred from the AC to the RC had a duty-limiting profile in the 24 months prior to separating from the AC, a rate on the order of 20 percentage points higher than the percentage who had a profile after affiliating with the RC. Those who spent at least 24 months in the RC had a higher profile rate than those who did not serve for 24 months. Among airmen, approximately 70 percent of all transfers had a duty-limiting profile before separating from the AC. We also looked at the percentage of transfers whose AC profile was 30 or more days in duration.[4] As the third row in Table 4.1 shows, approximately half of the transfers who had any AC profile had one that lasts 30 days or more. Therefore, if we use the presence of a profile on the AC as a benchmark, it is not at all uncommon for service members to have a profile at some point while in service. (Half or more of this population did while in the AC.)[5]

Table 4.1 is set up to compare those who serve 24 or more months in the RC to those who do not. We mentioned above that the results suggest that those who leave in the first 24 months do not have a higher RC profile rate and, therefore, may not leave because of medical issues. We can also look at whether those who had an AC profile prior to separation are more or less likely to complete 24 months of RC service after transferring. We know from the first two rows of the table how many transfers did and did not have an AC profile prior to separation and the number who did and did not serve 24 months in

---

[4] Although this policy was not in place over the period covered by this study, DoDI 1332.45, *Retention Determinations for Non-deployable Service Members*, states that service members who are on a profile for 30 days or more are considered nondeployable (Department of Defense Instruction 1332.45, 2018b).

[5] We do not have a way to compare the presence of an AC profile among transfers to those who did not affiliate with the RC after separating from the AC. We were unable to receive profile data on all service members and instead had to provide a sample to the Army and Air Force, and since the focus of this study is on AC to RC transfers, we only received data on those who were observed to join the RC.

the RC.[6] Among soldiers, a slightly higher proportion of those with an AC profile served a full 24 months in the RC (72 percent compared to 68 percent). The reverse is true for airmen. Among airmen, 79 percent with an AC profile served 24 months in the RC, compared to 83 percent who did not have a profile. The results are similar when restricting to AC profiles of 30+ days (third row of Table 4.1).

Next, we focused on profiles that might be a better signal of medical issues that result in individual medical nonreadiness. When issuing a profile, the services designate whether the profile is permanent or temporary. Per AR 40-501, "A temporary profile is given if the condition is considered temporary, the correction or treatment of the condition is medically advisable, and correction usually will result in a higher physical capacity" (Army Regulation 40-501, 2017, p. 76). This includes "medical or surgical care or recovering from illness, injury, or surgery, will be managed with temporary physical profiles until they reach the point in their evaluation, recovery, or rehabilitation" (p. 78). All other profiles are considered permanent. AFI 10-203, *Duty Limiting Conditions*, states that "profiles are descriptions of transient or permanent limitations to functioning" (Air Force Instruction 10-203, 2014, p. 5) and are described using the PULHES system (defined below) in accordance with AFI 48-123 (Air Force Instruction 48-123, 2013) and the Medical Standards Directory.

DoDI 1332.45 specifies that service members are considered temporarily nondeployable if, among other reasons, they have temporary profiles or are in LIMDU status, and light duty is not to be reported as nondeployable unless the duration exceeds 30 days (Department of Defense Instruction 1332.45, 2018b, p. 9). We use this 30-day threshold in our analysis of profiles to identify those that are associated with medical nonreadiness or nondeployability. As Table 4.1 shows, 18.9 percent of soldiers and 14.4 percent of airmen who spent fewer than 24 months in the RC had a profile lasting longer than 30 days. Among

---

[6] While not shown, the number who did not have a profile is the difference between rows 1 and 2. For example, among Army transfers who served fewer than 24 months, 11,279 had an AC profile prior to separation, and 11,411 did not. Among those who served 24 months or more, 28,868 had an AC profile prior to separation, and the remaining 24,538 did not.

those who served at least 24 months in the RC, 28.7 of soldiers and 24.7 percent of airmen had profiles lasting 30 days or longer.

The Army also provided data on the level of a profile (1–4), which is based on the individual's functional capacity across six organs or body systems: physical capacity or stamina (P), upper extremities (U), lower extremities (L), hearing and ears (H), eyes (E), and psychiatric (S), collectively called PULHES. Each body system is given a numerical designator, where 1 indicates a high level of fitness in that area and 4 indicates a functional limitation. The level on a profile is based on the six PULHES scores, where "an individual having a numerical designation of '1' under all factors is considered to possess a high level of medical fitness" and "a profile serial containing one or more numerical designators of '4' indicates that the individual has one or more medical conditions or physical defects of such severity that performance of military duty must be drastically limited" (AR40-501, p. 77).[7] Permanent profiles with a designation of 3 or 4 (P3 or P4) must be reviewed by a medical evaluation board physician or physical approval authority to assess whether the soldier meets medical retention standards.[8] We examined the P3/P4 profiles rate among AC to RC transfers and found that between 1 and 2 percent of transfers had a P3 or P4 profile after transferring to the RC.[9]

---

[7] We observe some Army profiles where all PULHES scores were '1' (i.e., 111111), but this is uncommon. When we do observe a PULHES of 111111, it appears to be a correction from a previous PUHLES, and some of these PULHES scores, indicating no impairment, are often accompanied by a comment explaining that it is a correction. Therefore, the profile records that we received from the Army appear to only be cases where the individual faces an impairment of level 2 or higher.

[8] If the physician determines that the soldier does not meet medical retention standards (as outlined in Army Regulation 40-501, 2017), the soldier must be referred to a medical evaluation board, the first step of DES. DES determines whether the soldier is fit to continue serving and may result in a medical discharge.

[9] The data we received from the Army indicated whether the profile was temporary or permanent and the designator on each PULHES factor (e.g., P1, U1, H2, L1, E1, S1). We used this information to determine the combined permanent-/temporary-level status of the profile (e.g., T1, T3, P2, P4).

Figure 4.1 explores the issue of months of RC service in more detail for Army transfers.[10] Table 4.1 showed that those who serve fewer than 24 months in the RC are no more likely to have a duty limitation than those who serve at least 24 months. We now condition on months served and examine the likelihood of serving 24 months for those with and without a profile. The first set of bars in Figure 4.1 shows that among transfers who serve at least 3 months in the RC, those with a profile are more likely to eventually serve for 24+ months than those without a profile by month 3 (69.79 percent compared with 66.51 percent, respectively). For the first year of RC service, this result is consistent with the table above: conditional on serving for a set number of months in the RC, those with a profile of 30+ days by that month are more likely to serve a full 2 years than those without a profile of 30+ days. However, after 12 months, the pattern changes. This change may be because the longer a transfer serves in the RC, the more likely they are to fulfill their MSO by the time they experience a medical condition that results in a duty limitation; therefore, they may choose to separate. For P3/P4 profiles, those who have a profile are consistently less likely to serve 24 months. Since P3/P4 profiles signal the most severe medical impairments, it is not surprising that those who have a P3/P4 profile are less likely to serve for 24 months in the RC. In Appendix B, Tables B.6 and B.7 show, at 3-month intervals, the number of transfers who do and do not have a profile and the number who do and do not serve 24 months. These numbers are the basis for the results in Figure 4.2.

Both the Air Force and Army provided information on the condition associated with the profile. The Air Force reported International Classification of Diseases, Ninth Revision (ICD-9), and ICD, Tenth Revision (ICD-10), codes, and the Army data contained a free-text field with a description of the profile.[11] We used the Clinical Clas-

---

[10] Because we only know the fiscal year of profile for Air Force transfers, we are unable to conduct this analysis because we cannot condition on the timing of profiles relative to RC affiliation month.

[11] ICD codes are published by the World Health Organization, and ICD is the diagnostic classification standard for clinical and research purposes, defining the universe of dis-

Figure 4.1
Percent Serving at Least 24 Months, by Profile Status and Months in the Reserve Component (Army)

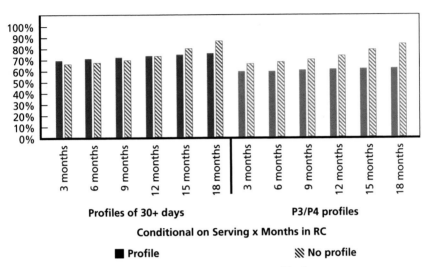

SOURCE: Author calculations using DMDC and Army profile data.
NOTE: The numbers represent the percentage of service members who serve at least x months who eventually serve at least 24 months in the RC. Percentages are shown separately for profiles of 30+ days and P3/P4 profiles and for those who do and do not have a profile as of month x.

sification Software (CCS) categories from the Agency for Healthcare Quality's Healthcare Cost and Utilization Project to aggregate individual ICD-9 codes into broader categories (the Healthcare Cost and Utilization Project's "Clinical Classifications Software (CCS) for ICD-9-CM" for diagnoses). We were able to directly map Air Force ICD-9 codes to CCS categories; for the Army, we manually coded free-text descriptions to ICD-9 codes, which we then converted to CCS categories.[12] Tables 4.2 and 4.3 show the top-ten most common conditions

---

eases, disorders, injuries, and other related health conditions (World Health Organization, undated).

[12] To do this, we tabulated the free-text descriptions and began working down the list, from the most frequent descriptions to the less common ones. We manually coded descriptions accounting for the most common 25 percent of conditions and then searched less frequent

for soldiers and airmen who had profiles lasting 30 or more days and, for soldiers, P3/P4 profiles.

The most common conditions listed on profiles were joint disorders, which were usually knee pain but also shoulder, ankle, and hip. (Among Air Force transfers, pregnancy and delivery was the most common condition listed on profiles.) Anxiety disorders were usually post-traumatic stress disorder (PTSD) and/or anxiety. Ear and sense organ disorders were usually listed as "hearing loss." Depression is a common condition appearing on profiles; it is categorized as a mood disorder. "Residual codes: unclassified" is nonspecific, but the description that comprises nearly all of that category is sleep apnea or, more generally, sleep disorder.

Timing of First Reserve Component Profile

Finally, profile data from the Army and Air Force allow us to measure how soon after joining the RC a transfer appears on a profile. While an individual can have multiple profiles during the 24 months we follow them in the RC, for simplicity of presentation, we focus on the first.

Months in Reserve Component at First Profile

Figures 4.2 and 4.3 show the distribution of months since affiliating with the RC at the time of the transfer's first profile for the Army, separately for those who serve at least 24 months in the RC and those who serve for fewer than 24 months.[13] The patterns are similar for the

---

descriptions for character words in the text string to identify other descriptions that meant the same thing and should be grouped together. For example, "hearing loss" was a common description that we manually coded, but the same condition, appearing less often as "Hearing Loss—H2" also appeared, and we grouped with the more common label.

[13] Because the Air Force profile data do not show start and end dates of the profiles, only duration, we are unable to do this analysis for the Air Force. There are a small number of transfers who appear on profile for the first time in month 24 even among those we have categorized as serving in SELRES for fewer than 24 months. This is because the data come from two sources. Our categorization for whether someone serves for 24 months comes from DMDC personnel files. Profile data come from the Army and Air Force and cover all components. There are several reasons why this discrepancy may occur. DMDC dates are always listed as the first of the month, which results in some lost precision about when someone affiliates with or leaves the RC, so it is possible someone actually serves for 24 months and

Table 4.2
Top-Ten Reserve Component Profile Conditions for Army Transfers

Profile Condition (CCS)	24 or More Months in the RC		Fewer Than 24 Months in the RC	
	Percent of Transfers with RC Profiles of 30+ Days (n = 15,342)	Percent of Transfers with RC P3/P4 Profiles (n = 715)	Percent of Transfers with RC Profiles of 30+ Days (n = 4,298)	Percent of Transfers with RC P3/P4 Profiles (n = 409)
Nontraumatic joint disorders (204)	26.64	43.50	26.50	20.84
Anxiety disorders (651)	16.95	36.78	15.43	18.74
Back problems (205)	15.92	40.98	16.91	22.10
Ear and sense organ disorders (94)	12.64	45.73	8.59	9.93
Miscellaneous mental health disorders (670)	7.87	3.78	5.49	1.12
Mood disorders (657)	7.23	12.17	7.14	7.69
Other pregnancy and delivery including normal (196)	5.19	—	4.82	—
Asthma (128)	4.82	18.88	5.40	9.37
Blindness and vision defects (89)	3.37	2.38	4.03	1.82
Residual codes: unclassified (259)	3.68	6.01	2.95	2.10
Diabetes mellitus without complication (49)	—	4.62	—	1.68

NOTES: Percent listed is of profiles with nonmissing conditions. A dash indicates there were not enough matches to report. (The smallest cell size we report on is ten.) Empty cells are those outside of the top ten for service members represented in that column. Numbers in parentheses following CCS descriptions are the CCS codes.

Table 4.3
Top-Ten Reserve Component Profile Conditions for Air Force Transfers

Profile Condition (CCS)	24 or More Months in the RC	Fewer Than 24 Months in the RC
	Percent of Transfers with RC Profiles Lasting 30+ Days (n = 5,741)	Percent of Transfers with RC Profiles Lasting 30+ Days (n = 827)
Other pregnancy and delivery including normal (196)	38.25	27.09
Nontraumatic joint disorders (204)	8.41	5.93
Back problems (205)	7.66	7.13
Disorders of teeth and jaw (136)	4.88	8.10
Other connective tissue disease (211)	5.09	4.35
Sprains and strains (232)	3.14	2.54
Residual codes: unclassified (259)	2.33	7.50
Anxiety Disorders (651)	2.11	3.87
Medical examination/ evaluation (256)	1.99	5.08
Ectopic pregnancy (180)	1.86	—
Other skin disorders (200)	—	2.30

NOTES: Percent listed is of profiles with nonmissing conditions. Empty cells are those outside of the top ten for service members represented in that column. Numbers in parentheses following CCS descriptions are the CCS codes.

two groups of soldiers: the largest percentage of transfers have their first profile beginning around the third to seventh month in the RC,

---

appears on a profile that month, but because of how dates are recorded, it appears as though they serve for 23. It is also possible that we categorize a transfer as serving fewer than 24 months if we observe them in SELRES for 23 months and they then return to the AC, where the Army or Air Force shows them on profile.

Figure 4.2
Distribution of Months of Reserve Component Service at the Time of First Profile of 30+ Days, Army Transfers

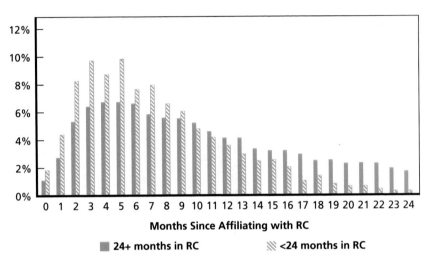

SOURCES: Date of RC affiliation from DMDC WEX file. Profile data from eProfile.
NOTE: Months of service calculated as the time between RC affiliation and start date of profile.

although the distribution is flatter for those who serve longer. Note that for those who serve fewer than 24 months in the RC, the population is declining over time in these figures.

With this understanding of RC profiles among AC to RC transfers, we now turn to information available at the time of AC separation that might be predictive of future RC medical issues that could lead to nondeployability.

Separation History and Physical Examination

DoD and VA have a joint program called the Separation Health Assessment (SHA) Program. The SHA is a single exam that can be conducted by DoD or VA around the time the service member is leaving active duty and serves the purposes of identifying medical conditions that are present at the time of separation, facilitating the transfer of care from

Figure 4.3
Distribution of Months of Reserve Component Service at the Time of First P3/P4 Profile, Army Transfers

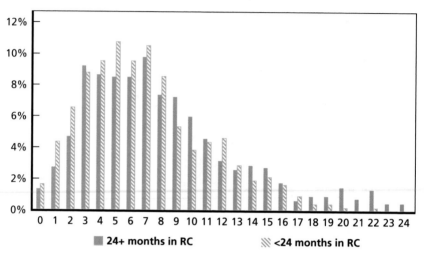

SOURCES: Date of RC affiliation from DMDC WEX file. Profile data from eProfile.
NOTE: Months of service calculated as the time between RC affiliation and start date of profile.

DoD to VA, and supports VA's evaluation of disability claims (Department of Defense Instruction 6040.46, 2016a; U.S. Department of Veterans Affairs, 2018). The department that conducts the assessment is based on timing and whether the service member is filing a disability claim with VA. When DoD conducts the SHA, it is called a SHPE.

According to DoDI 6040.46, a SHPE must be completed prior to separation at a military treatment facility or by DoD-contracted facilities for AC service members who are not filing a disability claim VA (Department of Defense Instruction 6040.46, 2016a). For those who are filing a claim with VA, the SHPE must be completed no later than 90 days prior to the date of separation from active duty, and the separation exam should be completed by VA when possible. DoDI 6040.46 also specifies that "service members who qualify for a SHPE because of retirement or separation are presumed fit for retention, except for those previously determined unfit and continued in a permanent limited duty status," and "if a condition is detected at the time of the

SHPE that would prevent a Service member from performing further duty if he or she were not separating, then the Service member will be referred for further evaluation and potential referral to the DES or IDES [Integrated Disability Evaluation System], in accordance with DODI 1332.18, or equivalent USCG [United States Coast Guard] process. Conditions that do not preclude completion of service but that require documentation of medical profiles for administrative purposes will be referred for such documentation according to Service specific procedures" (Department of Defense Instruction 6040.46, 2016a, p. 8). Documentation of a SHPE includes a code indicating that a SHPE took place, as well as "new diagnoses discovered in the SHPE not already listed on the Service member's automated problem list" (Department of Defense Instruction 6040.46, 2016a, p. 9).

Because the exam occurs so close to AC separation and therefore not long before affiliation with the RC, we received SHPE data from the Defense Health Agency (DHA) covering FY 2010–2018. As with profiles, we sent DHA a list of individuals in our AC to RC sample file, and they returned to us any patient encounter record containing the code indicating a SHPE had occurred (DOD0222) for those service members.[14] We used these data to assess whether medical issues identified during the SHPE differed for those who served fewer than 24 months in the RC versus those who stayed for at least 24 months and whether the SHPE contained information that would appear again after the service member affiliated with the RC.

Tables 4.4–4.7 show the number of transfers, by service, for whom we requested SHPE data and the number of individuals for whom we received data.[15] We also show the number and percentage of transfers who only had a code indicating that a SHPE occurred and the number whose record indicated some other medical condition. For

---

[14] These data come from DHA's Military Health System Data Repository Comprehensive Ambulatory/Professional Encounter Record enhanced file, which includes outpatient care delivered in DoD's military treatment facilities.

[15] Note that because the data began in FY 2010, we may be missing some SHPEs for the earliest cohorts of transfers, particularly those transfers who separated from AC in early 2010 and whose SHPE may have occurred in 2009.

those who had a condition other than DOD0222, we show the top-ten most common conditions identified during the SHPE. While our profile data is limited to only Army and Air Force transfers, DHA was able to supply SHPE data for transfers from all services.

We received SHPE data on nearly 60 percent of Army transfers, and among them, approximately half had a medical issue identified during a SHPE. Among Air Force transfers, slightly more than half had a SHPE done prior to AC separation, but only one-third of Air Force transfer SHPE records showed a medical condition. One-quarter of marine and one-third of sailor transfers had a SHPE record, and approximately 40 percent (for both services) reported a medical condition. Because DoD or VA can conduct the preseparation exam, we do not interpret the data to be missing if we do not observe a SHPE; instead, VA probably conducted an SHA for those service members without a SHPE. We do not have information on medical conditions that presented during VA exams, however, so the analyses that follow are more complete for the Army and Air Force.

There was not much variation in the conditions identified during the SHPE for those who ended up serving fewer than or at least 24 months in the RC. In other words, the conditions identified during the SHPE are not likely a reliable indicator of which service members will be able to complete 2 years of RC service following AC separation. Across services, nontraumatic joint disorders (including, for example, knee, shoulder, ankle, and hip pain) were common, as were back problems. Hearing and vision impairments were the most common conditions identified during SHPEs for sailors and marines; hearing defects were also on the top-ten list for soldiers and airmen, but vision defects were less common (at the bottom of the list for Army transfers and not one of the top-ten conditions for the Air Force). Codes indicating screening and history of mental health and substance abuse were the most common issue among Army transfers and were also part of the top-ten list for the other services.

The presence of a condition identified during a SHPE alone may not be an indicator of success in the RC, measured by time served, a duty limitation, or another negative medical readiness outcome, but it is possible that some conditions identified during a SHPE are more

Table 4.4
Separation History and Physical Examination Information for Army Transfers

	Fewer Than 24 Months in the RC	24+ Months in the RC
Number of transfers in sample	22,690	53,406
Number of transfers with SHPE (percent of transfers)	13,148 (57.95%)	30,665 (57.42%)
Only DOD0222	5,679 (43.19%)	12,925 (42.15%)
Other medical condition listed	7,469 (56.81%)	17,740 (57.85%)
Top-ten most common other conditions identified during SHPE (percent of other conditions listed)		
Ear and sense organ disorders (94)	18.89%	20.20%
Screening and history of mental health and substance abuse codes (663)	11.77%	15.68%
Administrative/social admission (255)	7.75%	6.70%
Nontraumatic joint disorders (204)	7.74%	7.56%
Screening for suspected conditions (not mental disorders or infectious disease) (258)	5.05%	3.45%
Back problems (205)	4.58%	4.42%
Residual codes: unclassified (259)	4.55%	3.92%
Disorders of lipid metabolism (53)	3.04%	3.00%
Medical examination/evaluation (256)	2.75%	—
Other nutritional, endocrine, and metabolic disorders (58)	2.60%	2.79%
Other connective tissue disease (211)	—	2.18%

NOTE: Numbers in parentheses following CCS descriptions are the CCS codes.

Table 4.5
Separation History and Physical Examination Information for Air Force Transfers

	Fewer Than 24 Months in the RC	24+ Months in the RC
Number of transfers in sample	5,750	23,285
Number of transfers with SHPE (percent of transfers)	3,136 (54.54%)	12,231 (52.53%)
Only DOD0222	1,948 (62.12%)	7,934 (64.87%)
Other medical condition listed	1,188 (37.88%)	4,297 (35.13%)
Top-ten most common other conditions identified during SHPE (percent of other conditions listed)		
Medical examination/evaluation (256)	15.53%	8.71%
Non-traumatic joint disorders (204)	9.19%	9.78%
Back problems (205)	5.66%	6.92%
Ear and sense organ disorders (94)	5.32%	6.46%
Administrative/social admission (255)	5.06%	3.14%
Other upper respiratory disease (134)	4.21%	4.43%
Residual codes: unclassified (259)	3.74%	2.98%
Other connective tissue disease (211)	3.28%	4.25%
Other nutritional, endocrine, and metabolic disorders (58)	3.28%	—
Screening and history of mental health and substance abuse codes (663)	3.06%	—
Other skin disorders (200)	—	3.82%
Blindness and vision defects (89)	—	3.00%

NOTE: Numbers in parentheses following CCS descriptions are the CCS codes.

Table 4.6
Separation History and Physical Examination Information for Marine Corps Transfers

	Fewer Than 24 Months in the RC	24+ Months in the RC
Number of transfers in sample	3,288	3,576
Number of transfers with SHPE (percent of transfers)	900 (27.37%)	1,056 (29.53%)
Only DOD0222 (percent of transfers with SHPE)	536 (59.56%)	618 (58.52%)
Other medical condition listed (percent of transfers with SHPE)	364 (40.44%)	438 (41.48%)
Top-ten most common other conditions identified during SHPE (percent of other conditions listed)		
Ear and sense organ disorders (94)	19.44%	22.02%
Blindness and vision defects (89)	14.34%	13.15%
Administrative/social admission (255)	7.51%	6.74%
Nontraumatic joint disorders (204)	7.24%	6.97%
Medical examination/evaluation (256)	6.57%	5.06%
Other Eye Disorders (91)	2.95%	3.60%
Back problems (205)	2.95%	4.49%
Screening for suspected conditions (not mental disorders of infectious disease) (258)	2.68%	—
Screening and history of mental health and substance abuse codes (663)	2.68%	2.58%
Residual codes: unclassified (259)	1.88%	—
Other nervous system disorders (95)	—	2.58%
Other connective tissue disease (211)	—	2.47%

NOTE: Numbers in parentheses following CCS descriptions are the CCS codes.

Table 4.7
Separation History and Physical Examination Information for Navy Transfers

	Fewer Than 24 Months in the RC	24+ Months in the RC
Number of transfers in sample	5,206	13,651
Number of transfers with SHPE (percent of transfers)	1,880 (36.11%)	4,790 (35.09%)
Only DOD0222	1,048 (55.74%)	2,593 (55.13%)
Other medical condition listed	832 (44.26%)	2,197 (45.87%)
Top-ten most common other conditions identified during SHPE (percent of other conditions listed)		
Ear and sense organ disorders (94)	19.75%	21.78%
Blindness and vision defects (89)	11.20%	9.63%
Administrative/social admission (255)	8.79%	6.65%
Medical examination/evaluation (256)	8.55%	3.70%
Nontraumatic joint disorders (204)	4.03%	4.72%
Back problems (205)	3.01%	3.18%
Disorders of lipid metabolism (53)	2.29%	3.27%
Other eye disorders (91)	2.29%	2.48%
Screening and history of mental health and substance abuse codes (663)	2.17%	2.23%
Other connective tissue disease (211)	2.17%	—
Other upper respiratory disease (134)	—	2.23%

NOTE: Numbers in parentheses following CCS descriptions are the CCS codes.

likely to match RC profile conditions that appear after the service member transfers. We explore this and potential matches between AC and RC profiles in the next section.

Active Component Indicators of Reserve Component Medical Nonreadiness

To assess whether information from a service member's time in the AC might be indicative of future RC medical readiness issues, we used SHPE data and AC profile information, along with profiles during the first 24 months of RC service, to see if conditions that limited AC duty or conditions identified just prior to separation presented once the service member affiliated with the RC.

The number of transfers whose RC profile condition matched a condition listed on a SHPE or an AC profile was, for some conditions, small, especially when we split the data into those transfers who served fewer than 24 months in the RC and those who served 24 or more months. In addition, for a given service, the most common SHPE conditions and profile conditions were similar for the two groups. Therefore, Tables 4.8 and 4.9 list the most common RC profile conditions for all transfers in a service, regardless of time served in the RC, and the percentage of transfers who had that same condition appearing on a SHPE or on an AC profile, for the Army and Air Force, respectively.

Because we know the duration of Army RC profiles and their permanent/temporary status, along with level (1–4), we report the match rate between AC profiles and SHPEs for both types of RC profiles. Table 4.8 shows that service members with RC profiles for nontraumatic joint disorder, hearing loss (ear and sense organ disorders), or back problems are the most likely to have had that condition on an AC profile or a SHPE, at a match rate as high as nearly 40 percent. For both types of profiles, the AC profile is generally more likely to match. Recall from Table 4.1 that the percentage of Army transfers with an RC P3/P4 profile was small, less than 2 percent, so cell sizes are small, in many cases too small to report match results.

Among Air Force transfers with an RC profile of 30+ days for a nontraumatic joint disorder, half had a matching AC profile. Other conditions with high match rates with AC profiles were back problems, other connective tissue disease, and sprains and strains. Like the Army, many cell sizes were too small to report in the SHPE match analysis; only one-third of Air Force transfers had a SHPE with a medical condition listed. In Appendix B, Tables B.8–B.13 show the number of transfers who do and do not have RC profiles (by type, 30+ days, and P3/P4) and the number who do and do not have AC conditions (on a profile or SHPE). The results presented in Tables B.8–B.13 were used to create the percentages in Tables 4.8 and 4.9.

While the results in Tables 4.8 and 4.9 suggest that medical issues known at the time of AC separation, either past profiles or findings from a SHPE, may help the services screen for medical issues that may persist after affiliation with the RC, it would also be possible to screen out too many potential transfers if medical conditions encountered during the service member's time in the AC did not appear on an RC profile. Tables 4.10 and 4.11 explore this possibility by showing the percentage of transfers who had a medical issue during the AC that showed up again on an RC profile. The results are shown separately for the Army and Air Force, by RC profile type, and by whether the condition appeared on an AC profile or during a SHPE. Indeed, for the conditions that appear most commonly on RC profiles, only a small proportion (less than 30 percent, but much lower for many conditions) of transfers who had those conditions on an AC profile or a SHPE later had an RC profile with that condition. For example, if the services were to screen on the presence of a nontraumatic joint disorder that appeared on an AC profile or during a SHPE, fewer than 12 percent of separating service members would have an RC profile with this condition in the first 24 months of RC service. Therefore, while information available at the time of AC separation may be useful for signaling something about a service member's potential fitness in the RC, that information should not be used in isolation because doing so could have a largely negative impact on recruiting transfers whose prior medical conditions would not persist.

Table 4.8
Match Rate Between Army Reserve Component Profiles and Active Component Profiles and Separation History and Physical Examination Conditions

Profile Condition (CCS)	RC Profiles of 30+ Days		RC P3/P4 Profiles	
	Percent of Transfers with Matching AC Profile Condition	Percent of Transfers with Matching SHPE Condition	Percent of Transfers with Matching AC Profile Condition	Percent of Transfers with Matching SHPE Condition
Nontraumatic joint disorders (204)	36.32	7.39	38.91	10.00
Anxiety disorders (651)	3.34	3.37	4.53	3.02
Back problems (205)	31.24	9.12	32.15	10.86
Ear and sense organ disorders (94)	9.49	16.46	18.09	15.58
Miscellaneous mental health disorders (670)	—	—	n/a	n/a
Mood disorders (657)	1.55	3.39	—	—
Other pregnancy and delivery including normal (196)	31.51	2.99	n/a	n/a
Asthma (128)	8.44	6.69	11.39	—
Blindness and vision defects (89)	—	2.32	—	—
Residual codes: unclassified (259)	6.66	8.97	—	—

NOTES: A dash indicates there were not enough matches to report (smallest cell size we report on is ten), and n/a indicates that this condition was not one of the ten most common among this type of profile. Diabetes mellitus without complication, other connective tissue disease, disorders of the teeth and jaw, and intracranial injury were part of the top-ten RC profile condition lists for some categories of Army transfers, but none had enough matches to report. Numbers in parentheses following CCS descriptions are the CCS codes.

Table 4.9

Match Rate Between Air Force Reserve Component Profiles and Active Component Profiles and Separation History and Physical Examination Conditions

Profile Condition (CCS)	RC Profiles of 30+ Days	
	Percent of Transfers with Matching AC Profile Condition	Percent of Transfers with Matching SHPE Condition
Other pregnancy and delivery including normal (196)	25.66	1.32
Nontraumatic joint disorders (204)	52.26	6.39
Back problems (205)	48.30	5.61
Disorders of teeth and jaw (136)	28.53	—
Other connective tissue disease (211)	41.16	3.66
Sprains and strains (232)	40.80	—
Residual codes: unclassified (259)	34.69	—
Anxiety disorders (651)	24.18	—
Medical examination/ evaluation (256)	—	—
Ectopic pregnancy (180)	26.67	—

NOTES: A dash indicates there were not enough matches to report (smallest cell size we report on is ten). Numbers in parentheses following CCS descriptions are the CCS codes.

Tables 4.10 and 4.11 were derived using Tables B.8–B.13. Tables B.8–B.13 can be used in other ways, including to calculate the percent of transfers who had an RC profile who did *not* have that condition on either an AC profile or a SHPE. In other words, how often does a medical condition appear shortly after joining the RC that did not appear while the service member was still serving in the AC? Tables 4.12 and 4.13 show these results, which in general are very small percentages. Taking Tables 4.10 and 4.11 together with 4.12 and 4.13, for

Table 4.10
Percentage of Service Members with Active Component Condition That Appears on a Reserve Component Profile, Army

Profile Condition (CCS)	RC Profiles of 30+ Days		RC P3/P4 Profiles	
	Percent of Transfers with Matching AC Profile Condition	Percent of Transfers with Matching SHPE Condition	Percent of Transfers with Matching AC Profile Condition	Percent of Transfers with Matching SHPE Condition
Nontraumatic joint disorders (204)	11.70	11.11	1.10	1.32
Anxiety disorders (651)	18.44	17.30	3.05	1.89
Back problems (205)	13.13	11.53	1.92	1.96
Ear and sense organ disorders (94)	21.04	3.39	6.92	0.55
Miscellaneous mental health disorders (670)	—	—	—	—
Mood disorders (657)	5.98	10.96	—	—
Other pregnancy and delivery including normal (196)	16.86	14.85	—	—
Asthma (128)	26.45	25.39	7.42	4.30
Blindness and vision defects (89)	—	2.23	—	—
Residual codes: unclassified (259)	10.48	2.75	—	—

NOTES: A dash indicates there were not enough nonmatches (transfers who had an AC condition but did not have an RC profile with the same condition) to report (smallest cell size we report on is ten). Numbers in parentheses following CCS descriptions are the CCS codes.

Table 4.11
Percentage of Service Members with Active Component Condition That Appears on a Reserve Component Profile, Air Force

Profile Condition (CCS)	RC Profiles of 30+ Days	
	Percent of Transfers with Matching AC Profile Condition	Percent of Transfers with Matching SHPE Condition
Other pregnancy and delivery including normal (196)	61.24	23.88
Nontraumatic joint disorders (204)	13.89	3.95
Back problems (205)	19.19	4.29
Disorders of teeth and jaw (136)	9.38	—
Other connective tissue disease (211)	11.62	3.02
Sprains and strains (232)	6.05	—
Residual codes: unclassified (259)	15.04	—
Anxiety Disorders (651)	25.17	—
Medical Examination/ Evaluation (256)	—	—
Ectopic Pregnancy (180)	11.23	—

NOTES: A dash indicates there were not enough nonmatches (transfers who had an AC condition but did not have an RC profile with the same condition) to report (smallest cell size we report on is ten). Numbers in parentheses following CCS descriptions are the CCS codes.

all conditions, we see that it is more common for transfers to have an AC condition that later shows up as an RC profile than it is for transfers to have an RC profile that does not match an AC profile or a SHPE condition. This suggests that information available at the time of AC separation may be useful in identifying potential future RC medical issues, and it is relatively less common for new issues to appear after the service members affiliates with the RC, but the rates are quite low in both cases.

Table 4.12
Percentage of Service Members with a Reserve Component Profile Who Did
Not Have an Active Component Profile or Separation History and Physical
Examination for the Same Condition, Army

Profile Condition (CCS)	RC Profiles of 30+ Days		RC P3/P4 Profiles	
	Percent of Transfers with No Matching AC Profile Condition	Percent of Transfers with No Matching SHPE Condition	Percent of Transfers with No Matching AC Profile Condition	Percent of Transfers with No Matching SHPE Condition
Nontraumatic joint disorders (204)	5.56	6.66	0.47	0.57
Anxiety disorders (651)	4.18	4.18	0.50	0.51
Back problems (205)	3.18	3.91	0.45	0.55
Ear and sense organ disorders (94)	2.78	2.97	0.43	0.52
Miscellaneous mental health disorders (670)	—	—	—	—
Mood disorders (657)	1.84	1.81	—	—
Other pregnancy and delivery including normal (196)	0.93	1.28	—	—
Asthma (128)	1.17	1.19	—	—
Blindness and vision defects (89)	—	0.89	—	—
Residual codes: unclassified (259)	0.85	0.85	—	—

NOTES: A dash indicates there were not enough nonmatches (transfers who had
an AC condition but did not have an RC profile with the same condition) to report
(smallest cell size we report on is ten). Numbers in parentheses following CCS
descriptions are the CCS codes.

Table 4.13
Percentage of Service Members with a Reserve Component Profile Who Did
Not Have an Active Component Profile or Separation History and Physical
Examination for the Same Condition, Air Force

	RC Profiles of 30+ Days	
Profile Condition (CCS)	Percent of Transfers with Matching AC Profile Condition	Percent of Transfers with Matching SHPE Condition
Other pregnancy and delivery including normal (196)	6.42	8.26
Nontraumatic joint disorders (204)	0.94	1.77
Back problems (205)	0.93	1.66
Disorders of teeth and jaw (136)	0.89	—
Other connective tissue disease (211)	0.69	1.10
Sprains and strains (232)	0.43	—
Residual codes: unclassified (259)	0.45	—
Anxiety disorders (651)	0.40	—
Medical examination/ evaluation (256)	—	—
Ectopic pregnancy (180)	0.31	—

NOTES: A dash indicates there were not enough nonmatches (transfers who had
an AC condition but did not have an RC profile with the same condition) to report
(smallest cell size we report on is ten). Numbers in parentheses following CCS
descriptions are the CCS codes.

Other Reserve Component Outcomes

Up to this point, this chapter has focused on duty limitations in the
RC and on information available before AC separation that might have
hinted the service member would have medical issues after transferring. We conclude this analysis by reporting other outcomes that may
occur after the service member affiliates with the RC, including being

placed on medical orders, receipt of incapacitation pay, disability evaluation, and discharge from the RC.

We consider the first three together. They represent more extreme outcomes (e.g., resulting in medical discharge) or take time to process and may occur after our observation window ends.

Incapacitation Pay

RC service members who are unable to perform their military duties or who are able to perform military duties but who are experiencing "a loss of earned income from non-military employment or self-employment as a result of" a condition that occurred while serving are entitled to incapacitation pay (Department of Defense Instruction 1241.01, 2016b, p. 14). To receive incapacitation pay, the injury, illness, or disease that limits military duties or results in lost income from nonmilitary or self-employment must have been incurred or aggravated in the line of duty, established in a line-of-duty determination (Department of Defense Instruction 1241.01, 2016b). To identify transfers who received incapacitation pay during their first 24 months in the RC, we requested an extract from DMDC's reserve pay file. The reserve pay file contains a variable indicating the amount of incapacitation pay the service member receives in a month.[16]

Medical Orders

10 U.S. Code 12301 outlines when military units or members of a reserve component of the U.S. military may be called to active duty. Section 12301(h) specifies:

1. When authorized by the Secretary of Defense, the Secretary of a military department may, with the consent of the member, order a member of a reserve component to active duty—
   a. to receive authorized medical care;

---

[16] We have reserve pay data through February 2017. That means we do not have 24 months of data for transfers who separated from AC in 2015 and 2016. Because of the low rates of receipt of incapacitation pay reported below, we do not think this censoring substantively affects our results.

      b. to be medically evaluated for disability or other purposes; or

      c. complete a required Department of Defense health care study, which may include an associated medical evaluation of the member.

Similarly, 10 U.S. Code 12322 states, "A member of a uniformed service described in paragraph (1)(B) or (2)(B) of section 1074a(a) of this title may be ordered to active duty, and a member of a uniformed service described in paragraph (1)(A) or (2)(A) of such section may be continued on active duty, for a period of more than 30 days while the member is being treated for (or recovering from) an injury, illness, or disease incurred or aggravated in the line of duty as described in any of such paragraphs."

AC to RC transfers who are called to active duty for medical treatment, recovery, evaluation for disability, or to complete a health care study are on medical orders, as defined by these two sections of U.S. code. Placement on medical orders soon after affiliating with the RC would be a cause for concern as part of DoD IG's observation about nondeployability among AC to RC transfers. To assess the rate at which transfers are placed on medical orders within their first 24 months of RC service, we requested data from DMDC's RC active service file. The RC active service file contains a statute code variable, which includes values indicating placement on medical orders (K = Section 12301(h) and R = Section 12322).

Disability Evaluation System

Service members who become injured or ill while serving and whose medical condition(s) may render them unfit for continued duty may be referred to DES. The outcome of the evaluation is whether or not the service member is fit for duty; those who are fit are returned to duty, and those who are not are either medically retired or separated with a disability rating and usually compensation and health care benefits. Referral to DES typically follows a period of treatment (usually up to one year for a single condition) and, especially for members of the RCs, a line-of-duty determination. We requested data from the Veterans Tracking Application (VTA), a system that records transactions that occur as service members are evaluated in DES. VTA is the official system of record for cases occurring under the Integrated Disability Evaluation System (IDES), a joint evaluation done by DoD and VA.[17] IDES began being rolled out across DoD in 2007 and was fully implemented by FY 2012. For the earliest cohorts of transfers (2010 and 2011), we may miss disability evaluations if the service member was referred to DES soon after affiliating with the RC and/or if the military treatment facility where treatment occurred was a late adopter of DES. In those cases, the evaluation may have been recorded in a service's system of record (and not in VTA), which we did not have access to for this study. Additionally, the VTA data used in our analysis were extracted in March 2017, so we will not observe disability evaluation outcomes that occur for some members of the 2015 and 2016 cohorts. Because of the censoring at the beginning and end of our data period, we are likely underreporting the rate of disability evaluations.

Table 4.14 displays, for each service, the rate at which AC to RC transfers receive incapacitation pay, are placed on medical orders, or are referred to DES within 24 months of affiliating with the RC. For the first two measures, incapacitation pay and medical orders, there were too few (fewer than ten) marines and sailors to be able to report results. All of these outcomes occur very infrequently, less than 0.5 percent of transfers in each service for any of the measures.[18] The one exception is placement on medical orders for the Air Force, which was explained to us as a reflection of procedure (the Air Force places service members

---

[17] Prior to 2007 and at some locations throughout the rollout of IDES (depending on a service member's duty assignment), DoD and VA conducted separate disability evaluations.

[18] As a point of reference, we used numbers published by the Accession Medical Standards Analysis and Research Activity on the number of service members evaluated for disability in FY 2014, along with FY 2014 total force as reported in the Population Representation in the Military Services, to calculate the percentage of RC service members who were evaluated for disability (Accession Medical Standards Analysis and Research Activity, 2016; Center for Naval Analyses, 2014). In FY 2014, 0.86 percent of the DoD SELRES population underwent disability evaluation (by service, the rates are 1.14 percent for the Army, 0.33 percent for the Air Force, 0.34 percent for the Marine Corps, and 0.20 percent for the Navy). Therefore, the percentage of transfers with a referral to disability evaluation is lower than the SELRES population. (On a service-by-service basis, only the Navy transfer rate is higher than the overall Navy RC rate.)

Table 4.14
Receipt of Incapacitation Pay, Placement on Medical Orders, and Percent
with Disability Evaluation System Referral Within 24 Months of Reserve
Component Affiliation

	Percent Receiving Incapacitation Pay	Percent Placed on Medical Orders	Percent with DES Referral
Army	0.07	0.09	0.52
Air Force	0.04	3.14	0.11
Marine Corps	—	—	0.20
Navy	—	—	0.47

coming off of a mobilization on medical orders), not necessarily evidence of a higher rate of medical issues.

Finally, throughout this report we have shown results separately for transfers who serve 24 or more months in the RC and those who leave the RC before completing 24 months of service. The reason for this is because during our conversations with representatives from the services, we frequently heard that 24 months of RC service is seen as valuable and a successful outcome of receiving a transfer from the AC. We do not have reliable information on why someone leaves the RC within the first 24 months, but if we follow them in our data after their time in the RC, we are able to observe if they reappear in the AC or appear in the IRR.[19] These outcomes should be seen as positive because the service member is still actively serving or available to the military if needed. Table 4.15 shows, by service, the percentage of transfers who join the IRR or return to the AC after they leave the RC.

Among those who serve fewer than 24 months in the RC, less than 5 percent in each service return to the AC. However, a much larger percent join the IRR to complete their service obligation. The IRR rate ranges from 30 percent among soldiers to nearly 90 percent among marines. These rates are consistent with earlier results, specifically years of service at the time of transfer (see Appendix A, Figures

---

[19] We looked at separation codes for those who leave the RC in the first 24 months, but most are missing or listed as unknown. One-quarter are coded as expiration of term of service.

Table 4.15
Affiliation with Individual Ready Reserve or Active Component Following Reserve Component Affiliation

	Fewer Than 24 Months in the RC		24+ Months in the RC	
	Percent Joining IRR	Percent Returning to the AC	Percent Joining IRR	Percent Returning to the AC
Army	30.43	2.31	9.09	0.69
Air Force	57.06	2.97	10.07	0.65
Marine Corps	87.56	4.23	37.11	1.99
Navy	59.30	3.67	14.24	2.06

A.14–A.17 for service-specific distributions). While the largest proportion of transfers from all services leave the AC at year 4 or 5, the distribution among marines is the most stark, with very few transferring out of the AC in later years. That means most marines who complete fewer than 24 months in the RC have time remaining on their initial obligation, which they appear to finish serving in the IRR. The percentage of transfers who serve 24 or more months in the RC and then affiliate with the IRR ranges from 9 to 37 percent. The percent returning to the AC is also lower among those who spend more time in the RC compared to those who leave in the first 24 months.

Summary

The SHPE (or SHA) is an opportunity to assess a service member's health just prior to separation from the AC. Based on the SHPE data we received (for 60 percent of Army transfers, 50 percent of Air Force transfers, and 25–35 percent of Marine Corps and Navy transfers), nontraumatic joint disorders, back problems, hearing and vision impairments, and screening and history of mental health and substance abuse were the most common medical issues identified during a SHPE. Using profile data that we received for soldiers and airmen, we found some overlap between SHPE information and duty-limiting conditions that appeared after the transfer affiliated with the RC. In particular, those with RC profiles for nontraumatic joint disorders, hearing loss, back problems, and, for the Air Force, connective tissue disease and sprains and strains, were relatively more likely to have the same condition appear during the SHPE or on a profile from their time in the AC. Other common RC profile conditions, including anxiety or mood disorders, asthma, vision loss, and disorders of the teeth and jaw, did not match at a high rate to SHPE conditions or AC profiles. It is worth noting that we looked for one-to-one matches between conditions coded using CCS, but "screening and history of mental health and substance abuse codes" on a SHPE might in reality be an indicator of mental health problems that may persist into RC service, such as anxiety or mood disorders. An important caveat of this analysis is that while information available at the time of AC separation (conditions found on AC profiles or during a SHPE) sometimes matches RC profiles, only a small percentage of service members (30 percent or fewer, depending upon the condition) with profiles or issues identified during a SHPE later have RC profiles for the same condition. Therefore, medical issues that appeared in the AC should not be assumed persistent, and this information should be used in conjunction with other information about the service member's likely fitness for RC service. RC profiles tend to appear soon after the transfer affiliates with the reserves (approximately 40 percent in the first 8 months for those who serve for 24 or more months in the RC and 60–80 percent for those who leave earlier).

Duty limitations lasting 30 or more days are the most common indicator of medical issues among those we identified. Very small proportions of transfers receive incapacitation pay, are placed on medical orders, or are referred for disability evaluation. Among those who do not serve for 24 months in the RC, a large proportion join the IRR (ranging from 30 percent of soldiers and nearly 90 percent of marines), and a small fraction return to the AC.

CHAPTER FIVE

Conclusions and Recommendations

In this study, we reviewed DoD and service policies that govern who can transfer from the AC to the RC and what medical screening must be done. We then used individual-level personnel, duty limitation (profile), and SHPE, pay, and disability data to describe the characteristics, duty limitations, and other medical issues of service members who separated from the AC and affiliated with the RC (SELRES) (based on an AC separation date between FY 2010 and FY 2016). This chapter summarizes those findings, describes limitations of our analyses, and offers recommendations for how DoD and the services can improve on the outcomes observed by DoD IG.

Summary of Findings

Policies and Service Implementation

Whether a service member transferring between the AC and RC must meet DoD standards for appointment to the military depends on the length of the gap between active and reserve service. One DoD regulation applies to applicants transferring to an RC after a period of more than 12 months since the separation physical, and the other applies to applicants for whom more than 6 months have elapsed since discharge. Our analysis showed that the vast majority (more than 80 percent) of service members transferring between the AC and RC do so within 6 months, so the two DoDIs that define standards for appointment do not apply to this population.

The services have policies that guide AC to RC transfers. Some services use a six-month gap, similar to DoD. The Army and Air Force specify that the service member must meet retention standards to qualify for RC service, and the Navy performs a medical review, looking specifically for new or materially changed conditions since leaving active service.

Characteristics of Active Component to Reserve Component Transfers

Our empirical analysis showed that 20,000 to 25,000 service members transferred from the AC to the RC in each year, FY 2010–2016. Approximately 80 percent had a gap of 1 or 2 months between AC separation and RC affiliation. DoD-wide, 14 percent of transfers were officers, and 70 percent had 8 or fewer years of service at the time of transfer (40 percent had 5 or fewer years). Officers transferred later in their careers; marines, earliest. We followed transfers for up to 24 months after they affiliated with the RC; 60 percent served for at least 24 months.

Duty Limitations of Active Component to Reserve Component Transfers

We received duty-limitation data (profiles) for Army and Air Force transfers. Among soldiers in our sample, 50 percent had a profile while in the AC (unconditional on duration or permanent/temporary status). Approximately 30 percent had a profile in their first 24 months after affiliating with the RC; of those, 80 percent were temporary profiles lasting 30 or more days, and 4.5 percent were permanent profiles, level 3 or 4. The most common conditions listed on soldier RC profiles were nontraumatic joint (e.g., knee, hip, ankle, shoulder) disorders, anxiety (usually PTSD), back problems, hearing loss, and mood disorders. RC profiles generally appeared soon after RC affiliation, within the first 3 to 7 months and sooner among those who served fewer than 24 months in the RC.

Two-thirds of the Air Force transfers in our sample had an AC profile, and one-third had an RC profile in the first 24 months after affiliation. Among those with an RC profile, 75 percent were 30 or more days in duration. The most common RC profile among airmen was related to pregnancy, followed by nontraumatic joint disorders, back problems, teeth and jaw disorders, other connective tissue disorders, and sprains and strains.

For the most common RC profile conditions, we looked back at the conditions listed on AC profiles and those found during the service member's SHPE to see how common it was for the conditions to match. Among soldiers, approximately one-third of transfers with an RC profile of 30+ days indicating a nontraumatic joint disorder or back problem had the same condition listed on SHPE. Back problems, hearing disorders, and nontraumatic joint disorders on P3/P4 profiles also matched AC profile conditions for 20 to 30 percent of soldiers. Among airmen, AC and RC profile conditions frequently matched for nontraumatic joint disorders and back problems (50 percent each), connective tissue disorders, and sprains and strains (40 percent each).

Other more extreme indicators of medical problems, such as receipt of incapacitation pay, placement on medical orders, or disability evaluation, were uncommon among transfers.

Limitations

There are a few limitations of our analysis that are important to acknowledge. First, DoD IG observed that some service members transferring from the AC to the RC had medical conditions that limited their deployability or made them nondeployable. As a consequence, IMR rates decreased. DoDI 6025.19 defines six elements of IMR: PHA, deployment-limiting medical and dental conditions, dental assessment, immunization status, medical readiness and laboratory studies, and individual medical equipment. The analyses in this study focused on deployment-limiting medical conditions, just one component of IMR. We were able to report the percentage of Army and Air Force transfers who had a duty-limiting profile after affiliating with the RC, but this is not a complete accounting of reduced IMR as documented by DoD IG. Furthermore, we were unable to obtain LIMDU data for Navy and Marine Corps transfers, so our full analysis only covers airmen and soldiers.

Second, we relied on SHPE data to assess whether information available at the time of separation from the AC could help predict that a duty limitation may arise after RC affiliation. Not all service members have a SHPE done by DoD; those who intend to file a disability claim with VA are encouraged to have a SHA done by VA. Among the transfers in our sample, we observed SHPE data for 60 percent of soldiers, 50 percent of airmen, 30 percent of marines, and 35 percent of sailors. It is unclear how our results might differ if we had complete data, including exams performed by VA. It is possible that transfers who are filing a disability claim with VA and whose data is therefore not part of our analysis are more likely to have a duty limitation after affiliating with the RC. On the other hand, service policies do not preclude service members with a VA disability rating or pending compensation from affiliating with the RC, so we do not assume that there is a difference in the ability to perform one's job duties that correlates with which department (DoD or VA) performs the separation exam. In addition to not having SHPE information for all transfers, we may not have observed all medical conditions. SHPE policy states that documentation should include new diagnoses not already listed on the service member's "automated problem list" (Department of Defense Instruction 6040.46, 2016a, p. 9). We did not have visibility of automated problem lists, so our analysis likely underreports the extent to which information available during the AC could be helpful in identifying potential problems that may arise after RC affiliation.

The Army and the Navy offer service members transitioning from active duty to an RC a deferral from involuntary deployments to allow service members the opportunity to establish their civilian lives, including beginning a new career or attending school without being interrupted by a deployment. Since prior research has shown that medical conditions are often identified as a reservist prepares to mobilize (Brauner, Jackson, and Gayton 2012), it is possible that medical issues are underreported in the data utilized in this study for service members who defer deployments. We did not receive data on deferments and have no way to estimate how the completeness of the data we did receive would be affected by deferments.

Finally, it was beyond the scope of this study to conduct a predictive analysis of which AC to RC transfers would experience duty limitations after affiliating with the RC. We reported on the most common conditions listed on RC profiles, the most common conditions identified during the SHPE, and the match rate between RC profile conditions and either AC profile conditions or SHPE conditions. A multivariate analysis would have been able to examine which service member characteristics were associated with RC duty limitations and the timing of when those limitations present relative to RC affiliation.

Recommendations

Based on the results of our analysis, we identified four recommendations for how DoD and the services may revise policy and use information available during the AC with the goal of reducing the number of transfers who affiliate with the RC with medical conditions that limit their deployability.

DoD should require service members to meet retention standards in order to be able to affiliate with the RC. Currently, DoDI 1200.15, which defines procedures for assigning service members to the RC, does not specify any requirements related to health or medical readiness.[1] The two DoDIs that do define medical standards, DoDI 6130.03 (Department of Defense Instruction 6130.03, 2018a) and 1304.26 (Department of Defense Instruction 1304.26, 2015a), do not apply to the AC to RC transfer population discussed in DoD IG's report. Specifically, DoDI 6130.03, *Medical Standards for Appointment, Enlistment, or Induction in the Military Services*, exempts service members who join an RC within 12 months of the separation physical, and DoDI 1304.26, *Qualification Standards for Enlistment, Appointment, and Induction*, states that it does not apply to service members dis-

---

[1] Since this report was written, DoDI 1200.15 has been revised and now includes a section on medical and deployable status standards for accession into the ready reserve. Department of Defense Instruction 1200.15, rev. *Assignment to and Transfer Between Reserve Categories and Discharge from Reserve Status*, Washington, D.C.: U.S. Department of Defense, 2019.

charged from one component who are reenlisting with another within 6 months. This gap in instruction pertaining to the standards that must be met by the transfer population, as defined throughout this study, should be resolved by requiring these transfers to meet medical retention standards.[2]

Because all service members are required to undergo a health assessment prior to separation, **we recommend that medical retention standards be applied at the time of the SHPE.** The intent of this exam is to identify medical conditions that are present at the time of separation to ease the transfer of care from DoD to VA. Because of the timing of the exam, no more than 90 days prior to the date of separation, it is the best opportunity to conduct a full assessment of the service member's health and their ability to perform their duties in the RC if retained. The feasibility of implementing this recommendation depends on how well some challenges can be overcome.

First, service members who are filing a VA claim for disability may have their separation exam done by VA, and this exam may occur earlier, up to 180 days prior to the date of separation. VA may not be equipped to assess whether the service member meets medical retention standards, so it may not be possible to apply retention standards at this stage for all transfers. A DoD official is required to review the results of the VA exam, so it might be possible to identify during this review whether medical retention standards are met. Alternatively, in the same way service members filing a claim with VA are encouraged to have their separation exam done by VA, perhaps service members planning to transfer to the RC, which usually happens within the first couple of months after AC separation and should be known at the time of separation, should have their exam done by DoD. It may also be possible to create a checklist that corresponds to medical retention standards that could be completed by a VA examiner achieve the outcome

---

[2] Since the time this research was completed and the report written, DoDI 1200.15 has been revised. It now states that service members re-accessing into the ready reserve within 12 months of separation must be fully medically ready according to individual medical readiness standards, and that service members who do and who are not deployable must be approved for retention by the gaining service. Department of Defense Instruction 1200.15, rev., 2019.

of assessing retainability regardless of which department conducts the exam. It is also worth noting that using the SHPE (and possibly the SHA) to evaluate whether the service member meets medical retention standards may increase the work involved in conducting the exam, thereby requiring additional resources.

Third, **we recommend that, to the extent possible, the requirements for transferring from the AC to the RC and the information used to determine whether the service member meets these requirements should be standardized across services and components.** The DoD and service policies currently vary in terms of who the policies apply to (e.g., based on number of months since AC discharge), what standards are applied, and what information is reviewed. This may be especially important for service members who leave one service and transfer to the RC of another; there is some evidence of this happening, as shown in Figure 3.3. If DoD implements our first recommendation and requires that retention medical standards be met, that will become a minimum requirement for the services. Further, DoD has a Retention Medical Standards Working Group that, at the time of this writing, is developing a DoD-wide retention policy; currently, these standards exist only at the service level. If all transfers have to meet these standards, it becomes easier to track requirements and to transfer service members between components.

Finally, **we recommend that individuals who determine whether service members are permitted to transfer from the AC to the RC make full use of information available at the time of AC separation.** During our conversations with service representatives, we learned that some or all of the following information is available to the receiving unit: duty-limiting conditions, VA disability ratings, SHPE results, and medical records. Although we did not have access to VA disability ratings or medical records, we did find that a significant proportion of service members with a deployment-limiting RC profile (either P3/P4 or 30+ days in duration) had the same condition on an AC profile or identified during the SHPE.[3] The match rate was

---

[3]   As measured by the CCS code.

as high as 50 percent for nontraumatic joint disorders among airmen, and several RC profile conditions matched at rates of 20 percent or higher for both services. While information available during the AC is valuable and frequently matches subsequent RC profile conditions, the services would have to exercise caution to not screen out transfers who have these conditions in the AC but do not experience issues once they join the RC. We found that a small percentage (30 percent or fewer) of transfers who have AC profiles later have RC profiles for the same condition; however, it is even less common for brand-new RC conditions to appear that do not match a condition listed on an AC profile or SHPE. Therefore, AC profiles and SPHE results should be just two inputs to any decision that is made about a service member's potential fitness for RC service.

Some of these recommendations may require additional resources or further analysis to determine exactly how to screen transfers. The benefits in terms of improved readiness and deployability must be compared to the costs of additional screening, both administrative and the possibility of screening out potential transfers who would have been a ready part of the RC. Our analysis showed that one out of every four or five Army and Air Force transfers were temporarily nondeployable within 24 months of affiliating with the RC, the majority in the first several months of RC service. Defining the medical standards that must be met and utilizing information already available are the first steps to reducing the number of service members who transfer to the RC with conditions that limit their deployability.

Additional Results by Service

Figure A.1 shows the number of service members who transfer from the AC to the RC, separately for enlisted personnel and commissioned officers. Figures A.2–A.5 show the number of service members who transfer to the RC, by fiscal year and service.

Figure A.1
Number of Active Component to Reserve Component Transfers,
All Services, 2010–2016

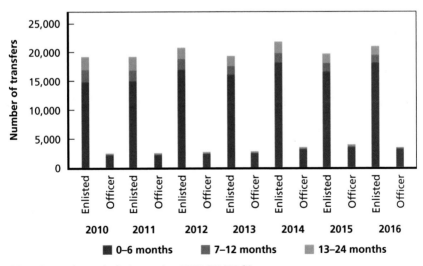

SOURCE: Author calculations using DMDC WEX file.
NOTES: Year is fiscal year of AC separation. Number of months represent months between AC separation and affiliation with SELRES.

Figure A.2
Number of Active Component to Reserve Component Transfers, Army,
2010–2016

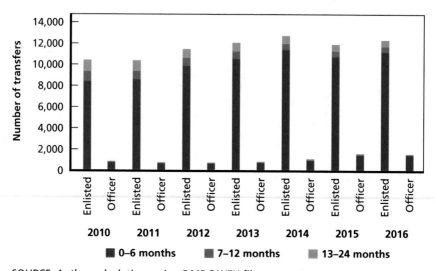

SOURCE: Author calculations using DMDC WEX file.
NOTES: Year is fiscal year of AC separation. Number of months represent months
between AC separation and affiliation with SELRES.

Figure A.3
Number of Active Component to Reserve Component Transfers, Air Force,
2010–2016

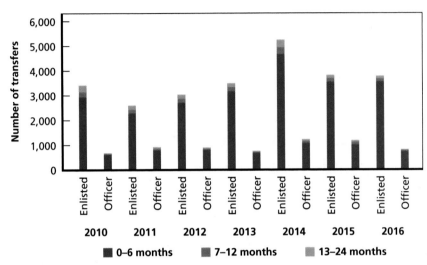

SOURCE: Author calculations using DMDC WEX file.
NOTES: Year is fiscal year of AC separation. Number of months represent months
between AC separation and affiliation with SELRES.

Figure A.4
Number of Active Component to Reserve Component Transfers, Marine Corps, 2010–2016

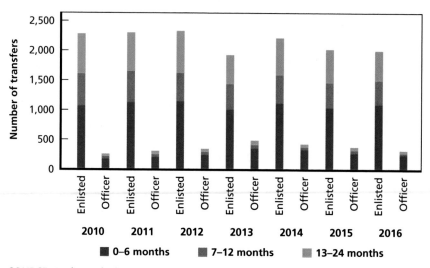

SOURCE: Author calculations using DMDC WEX file.
NOTES: Year is fiscal year of AC separation. Number of months represent months between AC separation and affiliation with SELRES.

Figure A.5
Number of Active Component to Reserve Component Transfers, Navy,
2010–2016

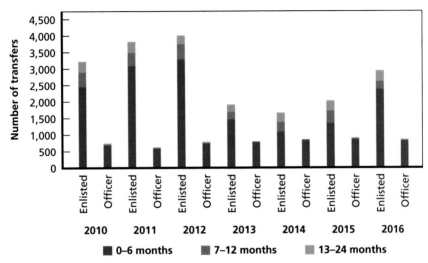

SOURCE: Author calculations using DMDC WEX file.
NOTES: Year is fiscal year of AC separation. Number of months represent months
between AC separation and affiliation with the RC.

Figures A.6–A.9 show the distribution of the gap between AC separation and RC affiliation, by service, for service members who transfer within six months.

Figure A.6
Distribution of Gap Between Active Component Separation and Reserve Component Affiliation, Army, 2010–2016

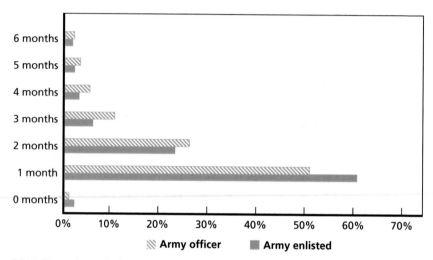

SOURCE: Author calculations using DMDC WEX filet.
NOTES: Number of months represent months between AC separation and affiliation with the RC.

Figure A.7
Distribution of Gap Between Active Component Separation and Reserve
Component Affiliation, Air Force, 2010–2016

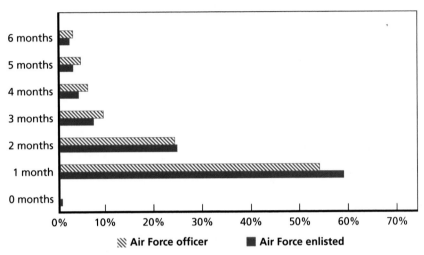

SOURCE: Author calculations using DMDC WEX file.
NOTES: Number of months represent months between AC separation and affiliation with the RC.

Figure A.8
Distribution of Gap Between Active Component Separation and Reserve Component Affiliation, Marine Corps, 2010–2016

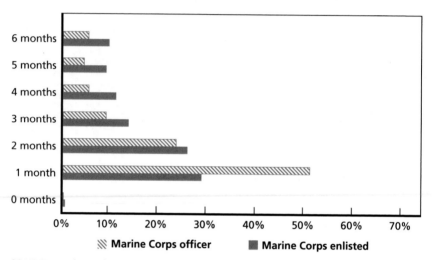

SOURCE: Author calculations using DMDC WEX file.
NOTES: Number of months represent months between AC separation and affiliation with the RC.

Figure A.9
Distribution of Gap Between Active Component Separation and Reserve Component Affiliation, Navy, 2010–2016

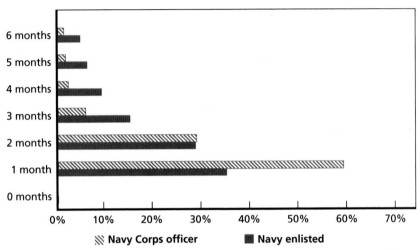

SOURCE: Author calculations using DMDC WEX file.
NOTES: Number of months represent months between AC separation and affiliation with the RC.

Figures A.10–A.13 show the pay grade distribution of AC to RC transfers for each service, where pay grade is measured at the time of RC affiliation.

Figure A.10
Pay Grade Distribution of Active Component to Reserve Component Transfers, Army, 2010–2016

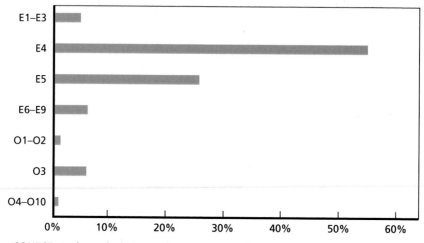

SOURCE: Author calculations using DMDC WEX file.

Figure A.11
Pay Grade Distribution of Active Component to Reserve Component
Transfers, Air Force, 2010–2016

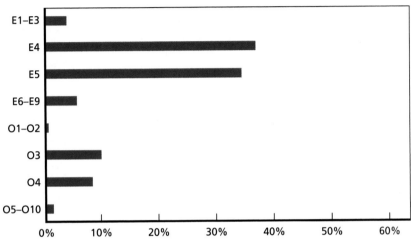

SOURCE: Author calculations using DMDC WEX file.

Figure A.12
Pay Grade Distribution of Active Component to Reserve Component Transfers, Marine Corps, 2010–2016

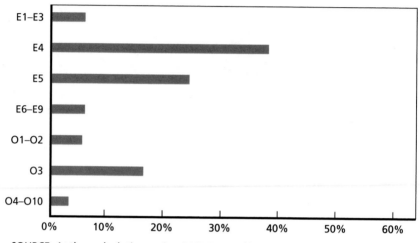

SOURCE: Author calculations using DMDC WEX file.

Figure A.13
**Pay Grade Distribution of Active Component to Reserve Component
Transfers, Navy, 2010–2016**

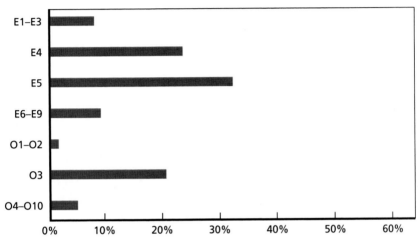

SOURCE: Author calculations using DMDC WEX file.

Figures A.14–A.17 show the years-of-service distribution among enlisted and commissioned officer AC to RC transfers. Each figure includes the overall years-of-service distribution for all transfers (as shown in Figure 3.5) for ease of comparison.

Figure A.14
Years-of-Service Distribution of Active Component to Reserve Component Transfers, Army and Overall, 2010–2016

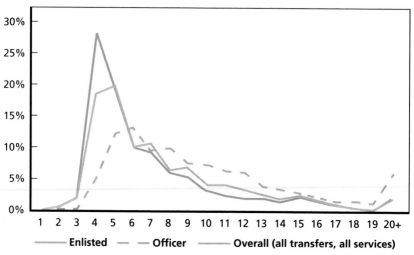

SOURCE: Author calculations using DMDC WEX file.
NOTES: The percentages do not add up to 100 because years of service was unavailable for some of the transfers in our sample. The overall line in this figure is the overall years-of-service distribution for all transfers from all services, both officer and enlisted. It is the same line that appears in Figure 3.5.

Figure A.15
Years-of-Service Distribution of Active Component to Reserve Component Transfers, Air Force and Overall, 2010–2016

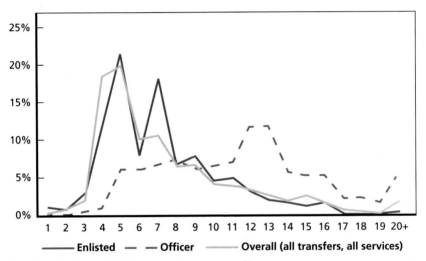

SOURCE: Author calculations using DMDC WEX file.
NOTES: The percentages do not add up to 100 because years of service was unavailable for some of the transfers in our sample. The overall line in this figure is the overall years-of-service distribution for all transfers from all services, both officer and enlisted. It is the same line that appears in Figure 3.5.

Figure A.16
Years-of-Service Distribution of Active Component to Reserve Component Transfers, Marine Corps and Overall, 2010–2016

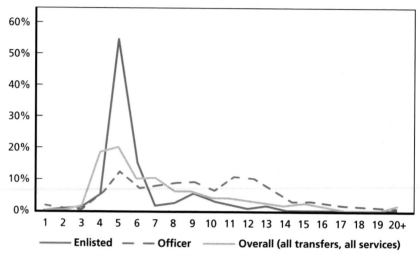

SOURCE: Author calculations using DMDC WEX file.
NOTES: The percentages do not add up to 100 because years of service was unavailable for some of the transfers in our sample. The overall line in this figure is the overall years-of-service distribution for all transfers from all services, both officer and enlisted. It is the same line that appears in Figure 3.5.

Figure A.17
Years-of-Service Distribution of Active Component to Reserve Component Transfers, Navy and Overall, 2010–2016

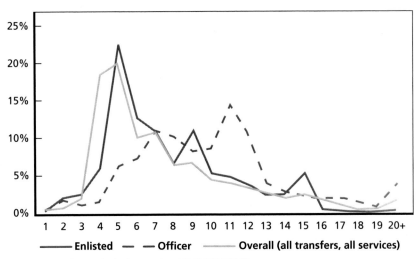

SOURCE: Author calculations using DMDC WEX file.
NOTES: The percentages do not add up to 100 because years of service was unavailable for some of the transfers in our sample. The overall line in this figure is the overall years-of-service distribution for all transfers from all services, both officer and enlisted. It is the same line that appears in Figure 3.5.

Additional Results

Table B.1
2013 Enlisted and Officer Selected Reserve End Strengths, by Service

	Reserves (SELRES)				National Guard (SELRES)			
	Enlisted	Commissioned Officer	Warrant Officer	Total	Enlisted	Commissioned Officer	Warrant Officer	Total
Army	162,959	32,000	3,250	198,209	312,670	36,601	8,464	357,735
Air Force	56,853	14,060	0	70,913	90,977	14,731	0	105,708
Marine Corps	35,401	3,821	279	39,501				
Navy	48,124	14,233	87	62,444				

SOURCE: Center for Naval Analyses, 2013.

Table B.2
2013 Enlisted and Officer Active Component End Strengths, by Service

	AC			
	Enlisted	Commissioned Officer	Warrant Officer	Total
Army	429,103	83,233	15,734	528,070
Air Force	261,775	64,798	0	326,573
Marine Corps	174,610	19,189	2,049	195,848
Navy	265,978	52,257	1,604	319,839

SOURCE: Center for Naval Analyses, 2013.

Table B.3
2013 Enlisted and Officer Prior Selected Reserve Prior Service Gains, by Service

	Reserves (SELRES)				National Guard (SELRES)			
	Enlisted	Commissioned Officer	Warrant Officer	Total	Enlisted	Commissioned Officer	Warrant Officer	Total
Army	13,960	4,554	319	18,833	14,041	3,769	710	18,520
Air Force	4,732	1,438	0	6,170	4,314	1,293	0	5,607
Marine Corps	3,207	1,046	46	4,299				
Navy	8,111	2,145	17	10,273				

SOURCE: Center for Naval Analyses, 2013.

Table B.4
Interservice Transfers as a Proportion of Active Component Separations,
2010–2016

		RC Service			
		Army	Air Force	Marine Corps	Navy
AC Service	Army	99.53%	0.42%	—	0.03%
	Air Force	0.76%	99.20%	—	0.04%
	Marine Corps	17.69%	5.18%	76.56%	0.57%
	Navy	2.17%	2.41%	—	95.41%

NOTES: The rows in this table add up to 100% and show, for transfers from a
particular AC service, what proportion transferred to each RC service. For example,
0.42 percent of transfers who served in the AC Army transferred to the Air Guard or
Air Reserve. Cells with a dash did not have a sufficient number of observations to
report.

Table B.5
Interservice Transfers as a Proportion of Reserve Component Gains,
2010–2016

		RC Service			
		Army	Air Force	Marine Corps	Navy
AC Service	Army	97.03%	1.09%	—	0.13%
	Air Force	0.28%	95.63%	—	0.06%
	Marine Corps	2.13%	1.64%	99.88%	0.28%
	Navy	0.56%	1.65%	—	99.54%

NOTES: The columns in this table add up to 100% and show, for transfers who
join a particular RC service, what proportion transferred from each AC service. For
example, 1.09 percent of service members who left Air Force AC service transferred
to the Army National Guard or Army Reserve. Cells with a dash did not have a
sufficient number of observations to report.

Table B.6
Number Serving at Least 24 Months, by Profile of 30+ Days Status and Months in Reserve Component (Army)

Conditional on Serving x Months	Profile of 30+ Days, Served 24+ Months	Profile of 30+ Days, Served <24 Months	No Profile of 30+ Days, Served 24+ Months	No Profile of 30+ Days, Served <24 Months
3 months	2,414	1,045	34,257	17,248
6 months	5,514	2,178	34,785	16,531
9 months	8,143	3,069	35,289	15,224
12 months	10,308	3,613	35,786	12,795
15 months	11,967	3,961	36,366	8.866
18 months	13,313	4,156	36,972	5,440

SOURCE: Author calculations using DMDC and Army profile data.
NOTE: The numbers represent the number of service members who serve at least x months who eventually serve at least 24 months in RC, separately for those who do and do not have a profile of 30+ days as of month x.

Table B.7
Number Serving at Least 24 Months, by P3/P4 Profile Status and Months in Reserve Component (Army)

Conditional on Serving x Months	Profile of 30+ Days, Served 24+ Months	Profile of 30+ Days, Served <24 Months	No Profile of 30+ Days, Served 24+ Months	No Profile of 30+ Days, Served <24 Months
3 months	130	88	36,541	18,205
6 months	314	210	39,985	18,499
9 months	489	310	42,943	17,983
12 months	588	363	45,506	16,045
15 months	648	392	47,685	12,435
18 months	673	405	49,612	9,191

SOURCE: Author calculations using DMDC and Army profile data.
NOTE: The numbers represent the number of service members who serve at least x months who eventually serve at least 24 months in RC, separately for those who do and do not have a P3/P4 profile as of month x.

Table B.8
Number of Transfers With and Without Active Component and Reserve Component Profile Conditions, RC Profiles of 30+ Days, Army

Profile Condition (CCS)	AC = 1, RC = 1	AC = 1, RC = 0	AC = 0, RC = 1	AC = 0, RC = 0
Nontraumatic joint disorders (204)	1,898	14,321	3,328	56,549
Anxiety disorders (651)	109	482	3,155	72,350
Back problems (205)	990	6,551	2,179	66,376
Ear and sense organ disorders (94)	219	822	2,089	72,966
Miscellaneous mental health disorders (670)	—	—	—	—
Mood disorders (657)	22	346	1,394	74,334
Other pregnancy and delivery including normal (196)	316	1,558	687	73,535
Asthma (128)	82	228	889	74,897
Blindness and vision defects (89)	—	—	—	—
Residual codes: unclassified (259)	46	393	645	75,012

NOTES: This table reports, for the top-ten most common RC profile conditions for Army transfers, the number of transfers who did and did not have that condition on an AC profile, paired with the number who did and did not have that condition on a RC profile of 30+ days. Rows with dashes are ones where one of the cells had fewer than ten transfers. Numbers in parentheses following CCS descriptions are the CCS codes. The numbers in this table were used to derive the numbers in the first column of Table 4.8.

Table B.9

Number of Transfers With and Without Active Component Separation History and Physical Examination and Reserve Component Profile Conditions, RC Profiles of 30+ Days, Army

Profile Condition (CCS)	AC = 1, RC = 1	AC = 1, RC = 0	AC = 0, RC = 1	AC = 0, RC = 0
Nontraumatic joint disorders (204)	386	3,088	4,840	67,782
Anxiety disorders (651)	110	526	3,154	72,306
Back problems (205)	289	2,217	2,880	70,710
Ear and sense organ disorders (94)	380	10,823	1,928	62,965
Miscellaneous mental health disorders (670)	—	—	—	—
Mood disorders (657)	48	390	1,368	74,290
Other pregnancy and delivery including normal (196)	30	172	973	74,921
Asthma (128)	65	191	906	74,934
Blindness and vision defects (89)	16	700	674	74,706
Residual codes: unclassified (259)	62	2,192	629	73,213

NOTES: This table reports, for the top-ten most common RC profile conditions for Army transfers, the number of transfers who did and did not have that condition on an AC SHPE, paired with the number who did and did not have that condition on an RC profile of 30+ days. Rows with dashes are ones where one of the cells had fewer than ten transfers. Numbers in parentheses following CCS descriptions are the CCS codes. The numbers in this table were used to derive the numbers in the second column of Table 4.8.

Table B.10
Number of Transfers With and Without Active Component and Reserve Component Profile Conditions, P3/P4 RC Profiles, Army

Profile Condition (CCS)	AC = 1, RC = 1	AC = 1, RC = 0	AC = 0, RC = 1	AC = 0, RC = 0
Nontraumatic joint disorders (204)	179	16,040	281	59,596
Anxiety disorders (651)	18	573	379	75,126
Back problems (205)	145	7,396	306	68,249
Ear and sense organ disorders (94)	72	969	326	74,729
Miscellaneous mental health disorders (670)	—	—	—	—
Mood disorders (657)	—	—	—	—
Other pregnancy and delivery including normal (196)	—	—	—	—
Asthma (128)	23	287	179	75,607
Blindness and vision defects (89)	—	—	—	—
Residual codes: unclassified (259)	—	—	—	—

NOTES: This table reports, for the top-ten most common RC profile conditions for Army transfers, the number of transfers who did and did not have that condition on an AC profile, paired with the number who did and did not have that condition on a P3/P4 RC profile. Rows with dashes are ones where one of the cells had fewer than ten transfers. Numbers in parentheses following CCS descriptions are the CCS codes. The numbers in this table were used to derive the numbers in the third column of Table 4.8.

Table B.11
Number of Transfers With and Without Active Component Separation History and Physical Examination and Reserve Component Profile Conditions, P3/P4 RC Profiles, Army

Profile Condition (CCS)	AC = 1, RC = 1	AC = 1, RC = 0	AC = 0, RC = 1	AC = 0, RC = 0
Nontraumatic joint disorders (204)	46	3,428	414	72,208
Anxiety disorders (651)	12	624	385	75,075
Back problems (205)	49	2,457	402	73,188
Ear and sense organ disorders (94)	62	11,141	336	64,557
Miscellaneous mental health disorders (670)	—	—	—	—
Mood disorders (657)	—	—	—	—
Other pregnancy and delivery including normal (196)	—	—	—	—
Asthma (128)	11	245	191	75,649
Blindness and vision defects (89)	—	—	—	—
Residual codes: unclassified (259)	—	—	—	—

NOTES: This table reports, for the top-ten most common RC profile conditions for Army transfers, the number of transfers who did and did not have that condition on an AC SHPE, paired with the number who did and did not have that condition on a P3/P4 RC profile. Rows with dashes are ones where one of the cells had fewer than ten transfers. Numbers in parentheses following CCS descriptions are the CCS codes. The numbers in this table were used to derive the numbers in the fourth column of Table 4.8.

Table B.12
Number of Transfers With and Without Active Component and Reserve Component Profile Conditions, RC Profiles of 30+ Days, Air Force

Profile Condition (CCS)	AC = 1, RC = 1	AC = 1, RC = 0	AC = 0, RC = 1	AC = 0, RC = 0
Other pregnancy and delivery including normal (196)	621	393	1,799	26,222
Nontraumatic joint disorders (204)	278	1,724	254	26,779
Back problems (205)	241	1,015	258	27,521
Disorders of teeth and jaw (136)	99	956	248	27,732
Other connective tissue disease (211)	135	1,027	193	27,680
Sprains and strains (232)	82	1,274	119	27,560
Residual codes: unclassified (259)	68	384	128	28,455
Anxiety disorders (651)	37	110	116	28,772
Medical examination/ evaluation (256)	—	—	—	—
Ectopic pregnancy (180)	32	253	88	28,662

NOTE: This table reports, for the top-10 most common RC profile conditions for Air Force transfers, the number of transfers who did and did not have that condition on an AC profile, paired with the number who did and did not have that condition on a 30+-day RC profile. Rows with dashes "-" are ones where one of the cells had fewer than 10 transfers. Numbers in parentheses, following CCS descriptions, are the CCS codes. The numbers in this table were used to derive the numbers in the first column of Table 4.9.

Table B.13
Number of Transfers With and Without Active Component Separation History and Physical Examination and Reserve Component Profile Conditions, RC Profiles of 30+ Days, Air Force

Profile Condition (CCS)	AC = 1, RC = 1	AC = 1, RC = 0	AC = 0, RC = 1	AC = 0, RC = 0
Other pregnancy and delivery including normal (196)	32	102	2,388	26,513
Nontraumatic joint disorders (204)	34	826	498	27,677
Back problems (205)	28	625	471	27,911
Disorders of teeth and jaw (136)	—	—	—	—
Other connective tissue disease (211)	12	386	316	28,321
Sprains and strains (232)	—	—	—	—
Residual codes: unclassified (259)	—	—	—	—
Anxiety disorders (651)	—	—	—	—
Medical examination/ evaluation (256)	—	—	—	—
Ectopic pregnancy (180)	—	—	—	—

NOTES: This table reports, for the top-ten most common RC profile conditions for Air Force transfers, the number of transfers who did and did not have that condition on an AC SHPE, paired with the number who did and did not have that condition on an RC profile of 30+ days. Rows with dashes are ones where one of the cells had fewer than ten transfers. Numbers in parentheses following CCS descriptions are the CCS codes. The numbers in this table were used to derive the numbers in the second column of Table 4.9.

References

10 U.S.C. 651, Members: Required Service, 2017. As of April 14, 2019:
https://www.law.cornell.edu/uscode/text/10/651

10 U.S.C. 1142, Preseparation Counseling; Transmittal of Certain Records to Department of Veterans Affairs, 2017. As of May 27, 2019:
https://www.law.cornell.edu/uscode/text/10/1142

10 U.S.C. 10145, Ready Reserve: Placement, 2017. As of April 14, 2019:
https://www.law.cornell.edu/uscode/text/10/10145

10 U.S.C. 12301, Reserve Components Generally, 2017. As of April 14, 2019:
https://www.law.cornell.edu/uscode/text/10/12301

Accession Medical Standards Analysis and Research Activity, *Disability Evaluation Systems Analysis and Research: Annual Report 2015*, Silver Spring, Md.: U.S. Department of Defense, March 11, 2016. As of April 14, 2019:
https://apps.dtic.mil/dtic/tr/fulltext/u2/1005611.pdf

Air Force Instruction 48-123, *Medical Examinations and Standards*, Washington, D.C.: U.S. Department of the Air Force, September 24, 2009. As of April 14, 2019:
https://www.seymourjohnson.af.mil/Portals/105/Documents/MDG%20Docs/AFI%2048-123.pdf?ver=2016-02-17-110439-997

Air Force Instruction 48-123, *Medical Examinations and Standards*, Washington, D.C.: U.S. Department of the Air Force, November 5, 2013, with changes through January 28, 2018. As of April 14, 2019:
http://static.e-publishing.af.mil/production/1/af_sg/publication/afi48-123/afi48-123.pdf

Air Force Instruction 10-203, *Duty Limiting Conditions*, Washington, D.C.: U.S. Department of the Air Force, November 20, 2014. As of April 14, 2019:
https://www.afpc.af.mil/Portals/70/documents/06_CAREER%20MANAGEMENT/03_Fitness%20Program/Air%20Force%20Instruction%2010-203.pdf?ver=2018-08-22-115744-620

Air National Guard, *Continue Your Air Force Career Part-Time in the Air National Guard*, undated. As of September 24, 2019:
https://www.goang.com/content/dam/goang/en/brochures/13066-ang_pcpf_brochure_low.pdf

Army Regulation 601-280, *Army Retention Program*, Washington, D.C.: U.S. Department of the Army, January 31, 2006 (Rapid Action Revision date September 15, 2011). As of April 14, 2019:
http://www.armywriter.com/r601_280.pdf

Army Regulation 601-280, *Army Retention Program*, Washington, D.C.: U.S. Department of the Army, April 1, 2016. As of April 14, 2019:
https://armypubs.army.mil/epubs/DR_pubs/DR_a/pdf/web/r601_280.pdf

Army Regulation 40-501, *Medical Services: Standards of Medical Fitness*, Washington, D.C.: U.S. Department of the Army, June 14, 2017.

Brauner, Marygail K., Timothy Jackson, and Elizabeth Gayton, *Medical Readiness of the Reserve Component*, Santa Monica, Calif.: RAND Corporation, MG-1105-OSD, 2012. As of April 14, 2019:
https://www.rand.org/pubs/monographs/MG1105.html

Center for Naval Analyses, "Population Representation in the Military Services," webpage, 2013. As of May 27, 2019:
https://www.cna.org/pop-rep/2013/contents/contents.html

———, "Population Representation in the Military Services, Fiscal Year 2014," webpage, 2014. As of April 14, 2019:
https://www.cna.org/pop-rep/2014/

Commander, Navy Recruiting Command Instruction 1130.8J, *Navy Recruiting Manual—Enlisted*, Millington, Tenn.: U.S. Department of the Navy, Policy and Programs Division, May 17, 2011. As of April 14, 2019:
http://navybmr.com/study%20material/CNRCINST%201130.8J%20(VOLUME-II).pdf

Commander, Navy Recruiting Command Instruction 1131.2F (undated), *Navy Recruiting Manual—Officer*, Millington, Tenn.: U.S. Department of the Navy, Policy and Programs Division, undated. As of November 18, 2016:
http://navybmr.com/study%20material/CNRCINST%201131.2F.pdf

Department of Defense Instruction 6490.07, *Deployment-Limiting Medical Conditions for Service Members and DoD Civilian Employees*, Washington, D.C.: U.S. Department of Defense, February 5, 2010. As of April 14, 2019:
https://www.esd.whs.mil/Portals/54/Documents/DD/issuances/dodi/649007p.pdf

Department of Defense Instruction 1235.13, *Administration and Management of the Individual Ready Reserve (IRR) and the Inactive National Guard (ING)*, Washington, D.C.: U.S. Department of Defense, October 18, 2013a. As of May 27, 2019:
https://www.esd.whs.mil/Portals/54/Documents/DD/issuances/dodi/123513p.pdf

Department of Defense Instruction 1304.25, *Fulfilling the Military Service Obligation*, Washington, D.C.: U.S. Department of Defense, October 31, 2013b. As of May 27, 2019:
https://www.esd.whs.mil/Portals/54/Documents/DD/issuances/dodi/130425p.pdf.

Department of Defense Instruction 1200.15, *Assignment to and Transfer Between Reserve Categories, Discharge from Reserve Status, Transfer to the Retired Reserve, and Notification of Eligibility for Retired Pay*, Washington, D.C.: U.S. Department of Defense, March 13, 2014a. As of April 14, 2019:
https://www.esd.whs.mil/Portals/54/Documents/DD/issuances/dodi/120015p.pdf

Department of Defense Instruction 6025.19, *Individual Medical Readiness (IMR)*, Washington, D.C.: U.S. Department of Defense, June 9, 2014b. As of April 14, 2019:
http://www.dtic.mil/whs/directives/corres/pdf/602519p.pdf

Department of Defense Instruction 1304.26, *Qualification Standards for Enlistment, Appointment, and Induction*, Washington, D.C.: U.S. Department of Defense, March 23, 2015a. As of April 14, 2019:
https://www.esd.whs.mil/Portals/54/Documents/DD/issuances/dodi/130426p.pdf?ver=2018-10-26-085822-050

Department of Defense Instruction 1215.13, *Ready Reserve Member Participation Policy*, Washington, D.C.: U.S. Department of Defense, May 5, 2015b. As of May 27, 2019:
https://www.esd.whs.mil/Portals/54/Documents/DD/issuances/dodi/121513p.pdf

Department of Defense Instruction 6040.46, *The Separation History and Physical Examination (SHPE) for the DoD Separation Health Assessment (SHA) Program*, Washington, D.C.: U.S. Department of Defense, April 14, 2016a. As of April 14, 2019:
https://www.esd.whs.mil/Portals/54/Documents/DD/issuances/dodi/604046p.pdf

Department of Defense Instruction 1241.01, *Reserve Component (RC) Line of Duty Determination for Medical and Dental Treatments and Incapacitation Pay Entitlements*, Washington, D.C.: U.S. Department of Defense, April 19, 2016b. As of April 14, 2019:
https://www.esd.whs.mil/Portals/54/Documents/DD/issuances/dodi/124101p.pdf

Department of Defense Instruction 6130.03, *Medical Standards for Appointment, Enlistment, or Induction into the Military Services*, Washington, D.C.: U.S. Department of Defense, May 6, 2018a. As of October 12, 2020:
https://www.esd.whs.mil/Portals/54/Documents/DD/issuances/dodi/613003p.pdf

Department of Defense Instruction 1332.45, *Retention Determinations for Non-deployable Service Members*, Washington, D.C.: U.S. Department of Defense, July 30, 2018b. As of April 14, 2019:
https://www.esd.whs.mil/Portals/54/Documents/DD/issuances/dodi/133245p.pdf?ver=2018-08-01-143025-053

Department of Defense Instruction 1200.15, rev. *Assignment to and Transfer Between Reserve Categories and Discharge from Reserve Status*, Washington, D.C.: U.S. Department of Defense, November 7, 2019. As of October 12, 2020:
https://www.esd.whs.mil/Portals/54/Documents/DD/issuances/dodi/120015p.pdf

Department of Defense Instruction 6130.03 V2, *Medical Standards for Military Service: Retention*, Washington, D.C.: U.S. Department of Defense, September 4, 2020. As of October 12, 2020:
https://www.esd.whs.mil/Portals/54/Documents/DD/issuances/dodi/613003v2p.pdf?ver=2020-09-04-120013-383

Healthcare Cost and Utilization Project, "Clinical Classifications Software (CCS) for ICD-9-CM," webpage, updated April 1, 2016, consulted April 26, 2016. As of April 14, 2019:
http://www.hcup-us.ahrq.gov/toolssoftware/ccs/ccs.jsp

Kotejin, Nicole, "Getting Out' Not So Fast," U.S. Army website, April 8, 2008. As of November 18, 2016:
https://www.army.mil/article/8351/Getting_Out__039__Not_so_Fast

Military Times, "2012 Insider's Guide to the Guard & Reserve," undated. As of May 27, 2019:
https://ec.militarytimes.com/guard-reserve-handbook/joining-up/status/

Myers, Meghann, "The Army Is Offering Two-Year Contracts and Cash Bonuses to Grow the Army," *Army Times*, February 19, 2017. As of May 27. 2019:
https://www.armytimes.com/news/your-army/2017/02/19/the-army-is-offering-two-year-contracts-and-cash-bonuses-to-grow-the-army/

Navy Bureau of Medicine and Surgery, *Change 166, Manual of the Medical Department*, Washington, D.C.: U.S. Department of the Navy, March 10, 2016.

Navy Personnel Command, *Enlisted Active Component to Reserve Component (AC2RC) Transition via the Career Transition Office (CTO)*, MILPERSMAN 1306-1501, November 30, 2017. As of May 27, 2019:
https://www.public.navy.mil/bupers-npc/reference/milpersman/1000/1300Assignment/Documents/1306-1501.pdf

Owens, Gabriel, "Reserves Offering Two-Year Deferment on Deployment," U.S. Navy website, March 14, 2007. As of November 18, 2016: http://www.navy.mil/submit/display.asp?story_id=28317http://www.public.navy.mil/bupers-npc/reference/messages/Documents/NAVADMINS/NAV2007/NAV07007.txt

Sanborn, James K., "Officers to Receive up to $20,000 to Join the Reserve," *Marine Times*, November 29, 2014. As of May 27, 2019: https://www.marinecorpstimes.com/news/your-marine-corps/2014/11/29/officers-to-receive-up-to-20000-to-join-the-reserve/

U.S. Department of Defense Inspector General, *Assessment of DoD-Provided Healthcare for Members of the United States Armed Forces Reserve Components*, Alexandria, Va.: U.S. Department of Defense, Report Number DODIG-2015-002, October 8, 2014. As of April 14, 2019: https://apps.dtic.mil/dtic/tr/fulltext/u2/a610379.pdf

U.S. Department of Veterans Affairs, *Separation Health Assessment for Service Members*, Washington, D.C.: U.S. Department of Veterans Affairs, August 2018. As of April 14, 2019: https://benefits.va.gov/BENEFITS/factsheets/serviceconnected/SeparationHealthAssessment.pdf

World Health Organization, "Classifications," webpage, undated. As of May 27, 2019: https://www.who.int/classifications/icd/en/